Surrogate Methods of Road Safety

This book provides a comprehensive understanding of accident data collection, analysis, and the use of surrogate safety measures (SSMs) from both vehicular and pedestrian perspectives. It discusses the application of simulation tools for surrogate safety analysis, with an emphasis on risk estimation and the integration of machine learning techniques. This book also explores the use of augmented and virtual reality for road user training and assessment, as well as safety concerns related to automated and connected vehicles. Field case studies offer a realistic view of on-site assessments, safety implications, and measures for safety enhancement.

- Explores surrogate safety methods in detail, including the identification of surrogate measures and their applications.
- Examines various SSMs, such as Post-Encroachment Time and Time-to-Collision, and identifies suitable SSMs for mixed traffic conditions, highlighting their strengths and weaknesses.
- Discusses international codes and standards at appropriate points in each chapter.
- Covers statistical methods, including Support Vector Machines, binary logit models, and ordered logit models, for estimating severity levels.
- Includes worked examples and numerical problems for practical understanding.

This book is intended for senior undergraduate and graduate students in road safety engineering, transportation, and civil engineering.

Surrogate Methods of Road Safety

K.V.R. Ravi Shankar and
Bhadradri Raghuram Kadali

CRC Press
Taylor & Francis Group
Boca Raton London New York

CRC Press is an imprint of the
Taylor & Francis Group, an **informa** business

Designed cover image: © K.V.R. Ravi Shankar and Bhadradri Raghuram Kadali

First edition published 2026
by CRC Press
2385 NW Executive Center Drive, Suite 320, Boca Raton FL 33431

and by CRC Press
4 Park Square, Milton Park, Abingdon, Oxon, OX14 4RN

CRC Press is an imprint of Taylor & Francis Group, LLC

ISBN: 9781032979793 (hbk)
ISBN: 9781032995472 (pbk)
ISBN: 9781003604761 (ebk)

DOI: 10.1201/9781003604761

Typeset in Times
by codeMantra

This book is dedicated to all the victims of the road who lost their valuable life in road accidents

Contents

Authors' Biography

K.V.R. Ravi Shankar is working as Associate Professor in the Department of Civil Engineering, National Institute of Technology (NIT) Warangal, Telangana, India. Ravi Shankar received his Ph.D. from the Indian Institute of Technology Bombay in 2011. He is having 20 years of experience in the field of teaching and research. His research interests center on innovative road safety strategies, machine learning application for traffic management and road safety, pedestrian behavioral analysis and modeling, and crowd behavior modeling. Ravi Shankar guided eight Ph.D. thesis works and four more are in progress. He also guided 64 M.Tech. project works. He is the author of about 90 publications in leading international journals and conferences. He was awarded the Young Researcher [Engineering] Award in 2020 by NIT Warangal Alumni Association. He is a recipient of research grants from SPARC, Department of Science and Technology, and Ministry of Road Transport and Highways and has a collaboration with Lund University, Sweden and UNSW, Australia. He worked on consultancy projects related to development of toolkits, traffic management planning, intersection design, and road safety improvement. He is a reviewer of several transportation-related journals and a prominent invited expert on road safety-related topics.

Bhadradri Raghuram Kadali is currently working as Assistant Professor in the Department of Civil Engineering, National Institute of Technology, Warangal (NITW). Raghuram obtained his Ph.D. degree from the Indian Institute of Technology Bombay in 2017 and Masters in Transportation Engineering from NITW. He has eight years of teaching and research experience. He has guided one Ph.D., 20 M.Tech. dissertations, five B.Tech. project works, and currently three research scholars are pursuing their Ph.D. Raghuram has successfully completed three externally funded R&D projects as principal investigator and associated with another two projects as co-principal investigator. He is currently working in traffic safety, pedestrian travel behavior, and transportation planning. He has published 50 papers in reputed journals and international conferences.

Preface

Road safety is one of the crucial challenges faced by the present road transportation sector across the world. At one or other point of time, everyone in this world was a mute victim to the ever-increasing burden of road accidents. Accident frequency and severity are direct measures of road safety and accident data are used to measure road safety. But accidents are rare events and sometimes collection of sufficient accident data is not possible. In the absence of accident data, surrogate safety measures (SSMs) are used to predict and analyze the frequency and severity levels of possible potential traffic conflicts using videography data. SSMs are indirect and complementary safety measures, and they don't rely on accident data. SSMs are more proactive, more accurate, more informative, and more time efficient. In the recent past, the use of SSMs has been widely spread across the world.

This book is intended as a one-stop reference for safe system concepts along with various SSMs adopted by various agencies across the world, data collection procedures, and methods of analysis. This book lays the foundation for the benefit of practicing road safety engineers, traffic engineers, and transportation engineers in adopting the safe system concepts for creating a safer road network. Through extensive real-world examples, case studies, and practical relevance, this book explains the intrinsic principles of safer system approach. Furthermore, pedestrian safety, use of simulation methods, and augmented and virtual reality considering the safe system concepts are further presented as a part of this book.

Acknowledgments

This book includes many ideas and concepts experienced and taught by the authors as an instructor and researchers on topics related to road safety. We would like to acknowledge the research scholars, students, faculty colleagues, and administration of National Institute of Technology Warangal, Telangana, India for their valuable support and encouragement. We are particularly thankful to all those who were courageous enough to allow to use draft sentences, paragraphs, or sections as examples or problems in this book. We are also thankful to our family members and friends for the encouragement provided at various stages of the book preparation.

We would like express our deepest gratitude to Prof. Aliaksei Laureshyn from Department of Technology and Society, LUND University, Sweden for the support and initiative towards introducing safe system approach concept. We would like to acknowledge the support of the international collaborative research project between NIT Warangal, India and Lund University, Sweden titled "Safe System perspective on vulnerable road users in India" (Project code: P3117) as a part of the SPARC (Scheme for Promotion of Academic and Research Collaboration) project fund sanctioned by Ministry of Education, Government of India, in the year 2024. We are also grateful to all the reviewers who have edited and commented on various draft chapters and the book proposal form. Finally, we would like to express appreciation to our beloved Director Prof. Bidyadhar Subudhi and everyone at the National Institute of Technology Warangal, Telangana, India, for their unwavering support and encouragement.

1 Introduction to Road Safety

1.1 CURRENT STATUS OF ROAD SAFETY

Road traffic safety involves the implementation of strategies and measures to prevent fatalities and serious injuries among individuals using roadways. Common road users include pedestrians, cyclists, drivers, vehicle occupants, horse riders, and passengers of on-road public transportation. Globally, approximately 1.19 million people lose their lives in road traffic accidents each year, and an additional 20–50 million suffer non-fatal injuries. Around 92% of the global fatalities occur in low- and middle-income countries, though these countries carry 60% of the world's vehicular population. Vulnerable road users, such as pedestrians, cyclists, and motorcyclists with their passengers, account for more than half of these accidents. Road traffic injuries are the leading cause of death for individuals aged between 16 and 29 years, particularly affecting males under 25 years, who represent 73% of road traffic deaths. Developing economies, particularly low- and middle-income countries, bear a disproportionate burden, contributing to 93% of fatalities. Beyond the human toll, road traffic injuries impose a significant economic burden, encompassing treatment costs for the injured and the loss of productivity for those killed or disabled. These injuries also have a substantial impact on national economies, causing countries to lose 3% of their annual gross domestic product. Recognizing the severity of the situation, established measures exist to mitigate the risk of road traffic injuries and fatalities. The 2030 Agenda for Sustainable Development has set ambitious targets to address and reduce road traffic injuries.

In India, road transportation emerges as the most cost-effective mode for both freight and passenger travel, especially given its extensive coverage in densely populated regions. The country confronts heightened exposure to challenging traffic conditions due to increased motorization and urbanization, driven by a robust economic growth rate. Consequently, the incidence of road accidents, traffic injuries, and fatalities remains notably elevated. Globally, road accidents stand out as a leading cause of death, with a significant concentration among individuals aged 15–49. In the year 2023 alone, road crashes in India resulted in the loss of approximately 1.72 lakh lives and caused injuries to over 4.62 lakh individuals. Given that road accidents stem from a complex interplay of various factors, a comprehensive set of measures is imperative to reduce both the number of accidents and resulting fatalities.

In the calendar year 2023, States and Union Territories (UTs) collectively reported a total of 4,80,583 road accidents. These incidents resulted in the loss of 1,72,890 lives and caused injuries to 4,62,825 individuals. Notably, there was a 4.2% increase in the number of road accidents in 2023 compared to the preceding year, 2022. Similarly,

DOI: 10.1201/9781003604761-1

fatalities and injuries stemming from road accidents also witnessed a rise, with a 2.6% and 4.4% increase, respectively. On average, this translates to 1,317 accidents and 474 deaths every day or 55 accidents and 20 deaths every hour in the country.

1.2 NEED FOR ROAD SAFETY MEASURES

In today's world of ever-increasing road accidents, road safety improvement plays a crucial role in safeguarding precious lives and burden on society. With an increasing vehicular population all over world, heavier traffic leads to a higher percentage and likelihood of accidents across countries. In fact, road accidents are the seventh leading cause of death in the world.

Most road accidents are caused by human error; although, sometimes unmaintained roads and horrible weather conditions lead to accidents, as well. The major factors that lead to traffic accidents are as follows: drunk driving, tiredness, speeding, road rage, failure to look or judge another car's speed properly, losing control of the vehicle, tailgating, distraction, carelessness, and inexperience.

There must be number of interventions required to arrest this abnormal increase in number of road accidents across the world. The government, non-governmental organizations, civilians, and more need to start initiatives toward more technical comprehensive safety improvement measures. Increased awareness of speed limits, alcohol limitations, seatbelt usage, child restraints, mobile usage, and construction zones needs to be emphasized. New policies need to be put in place to create safer streets, and older (but good) policies need to be enforced harder for them to take proper effect. Vehicles need to continue to evolve to be safe modes of transportation for not only the driver, but also for pedestrians and cyclists.

To lower the number of accidents, road traffic authorities need to assess problems with existing laws, rules, and systems to determine the real cause of accidents. A new and improved road safety strategy can then be put in place, which will have set actions to help prevent crashes from happening. In case of do-nothing scenario, as per the estimates of Association for Safe International predictions, traffic injuries will jump to the fifth leading cause of death by 2030 in the world.

1.3 REASONS FOR ROAD ACCIDENTS

Road traffic accident results from a combination of factors related to the components of the system comprising roads, the environment, vehicles and road users, and the way they interact. These can be grouped under the following heads as depicted in Table 1.1.

Some of these factors have a detrimental effect on the occurrence of road accidents. For example, every 1% increase in mean speed produces a 4% increase in the fatal crash risk and a 3% increase in the serious crash risk. Driving under the influence of alcohol and any psychoactive substance or drug increases the risk of a crash that results in death or serious injuries. In the case of drink-driving, the risk of a road

TABLE 1.1
Factors resulting in accidents

Road User Factors	Law and Enforcement	Road Factors	Vehicular Factors	Others
• Over-speeding • Distracted driving • Non-compliance of motorcycle helmets, seatbelts, and child restraints • Fatigue • Carelessness	• Inadequate speed limits • Need for national safety laws • Reliable enforcement strategies	• Pavement failures • Inadequate footpaths, crossings, intersection designs • Poor geometrical elements • Insufficient friction	• Unsafe vehicles • Improper tire pressure • Braking failure • Lack of adequate maintenance • Improper use of vehicle restraint system	• Inadequate post-crash care • Weather conditions like heavy rain, fog, and snow

traffic crash starts at low levels of blood alcohol concentration (BAC) and increases significantly when the driver's BAC is ≥0.2% (0.04 g/dl).

Correct use of helmet can lead to approximately 42% reduction in the risk of fatal injuries and 69% reduction in the risk of head injuries. Wearing a seatbelt reduces the risk of death among drivers and front seat occupants by 45–50%, and the risk of death and serious injuries among rear seat occupants by 25%. The use of child restraints can lead to a 60% reduction in deaths.

There are many types of distractions that can lead to impaired driving. The distraction caused by mobile phones is a growing concern for road safety. Drivers using mobile phones are approximately four times more likely to be involved in a crash than drivers not using a mobile phone. Using a phone while driving slows reaction times (notably braking reaction time, but also reaction to traffic signals), and makes it difficult to keep in the correct lane, and to keep the required following distances.

1.4 CAUSES AND EFFECT OF ROAD ACCIDENTS

Although road defects are not the major cause of road crashes, efforts to improve road design, construction, and maintenance are often highly cost-effective—and much easier than trying to improve the skills and attitudes of drivers. Research shows that there are three contributing factors to road crashes:

- Human factors (involved in about 95% of crashes)
- Road and road environment factors (involved in about 28% of crashes)
- Vehicle factors (involved in around 8% of crashes)

In the case of road accidents, multiple factors from the above three groups often combine in a chain of events that result in an accident. One such example can be depicted as follows:

> Driver is tired >> driver is on an unfamiliar road >> road is wet >> brakes on the car are poorly adjusted >> driver is travelling too fast for the conditions >> a sharp bend in the road is hidden by a vertical crest >> there is no sign warning of the sharp bend >> driver loses control, skids, runs off the road and crashes into a tree.

In this example, the road defects (poor alignment, lack of warning sign) are not the major cause of the accident, but they would have been relatively easy to avoid (through auditing)—and experience shows that it only takes one of the links in the chain to be broken to prevent an accident or reduce its consequences.

1.5 CONVENTIONAL AND SURROGATE APPROACH

In the traditional approach of traffic safety analysis, one needs to collect the historical accident record at a particular location and safety evaluation is done in terms of number of fatal, injury, and property damage accidents and their consequences. It is termed as a reactive approach of traffic safety analysis. As in this approach analysis was done using the historical accident data, it was also termed as after-the-fact analysis. Although conventional data provides basis for the modeling of various transportation planning schemes, there is uncertainty in quality and broad availability of data, lack of accuracy and reliability of the data, and dynamic randomness of the data. In the recent past, surrogate safety analysis techniques attempt to lessen the burden of data collection cycle by identifying specific surrogate indicators that are highly correlated to risk of collisions and occur much more frequently than crashes but might otherwise not be reported through the traditional channels. In contrast to traditional collision studies requiring many years of data, surrogate safety studies only require limited video footage from any kind of camera, making it a time and cost-effective method. In addition, the burden of road accidents on the road users and the nation is lessened through this pro-active approach.

1.6 ROAD SAFETY INITIATIVES BY GOVERNMENT OF INDIA

- The Government has approved a National Road Safety Policy. This policy outlines various policy measures such as promoting awareness, establishing road safety information database, and encouraging safer road infrastructure including application of intelligent transport and enforcement of safety laws.
- The Government has constituted the National Road Safety Council as the apex body to take policy decisions in matters of road safety.
- The Ministry has requested all States/UTs to constitute State Road Safety Council and District Road Safety Committees, and to hold their meetings regularly.
- The Ministry has formulated a multi-pronged strategy to address the issue of road safety based on 6'E's, viz. Education, Engineering (both of roads and vehicles), Enforcement, Emergency Care, Evaluation, and Encouragement.

- Road safety has been made an integral part of road design at the planning stage.
- Tightening of safety standards for vehicles like seatbelts and anti-lock braking system.
- The Ministry of Road Transport and Highways has introduced Motor Vehicle (Amendment) Bill, 2016, in Lok Sabha in August 2016 containing the different provisions of road safety. The Bill addresses road safety issues by providing for stricter penalties for traffic offences, permitting electronic and IT-enabled enforcement, improving fitness certification and licensing regime, statutory provision for protection of good Samaritans, etc.

1.7 ACCIDENT ANALYSIS

A road accident refers to any accident involving at least one road vehicle, occurring on a road open to public circulation, and in which at least one person is injured or killed or occurrence of a property damage. The problem of accident is very acute in highway transportation due to complex flow pattern of vehicular traffic, presence of mixed traffic along with pedestrians.

Road accidents cannot be totally prevented but, by suitable traffic engineering and management, the accident rate can be reduced to a certain extent. For this reason, systematic study of traffic accidents is required to be carried out. Proper investigation of the cause of accident will help to propose preventive measures in terms of design and control. Accident analysis is the major part in the accident studies. Accident analysis involves the tasks of accident data collection and analysis of collected data to suggest suitable safety improvement measures.

The accident data collection involves the data collection by the police or a road safety engineer. The data to be collected should comprise the following parameters:

- General: Date, time, person involved in accident, classification of accident like fatal, serious, or minor
- Location: Description and detail of location of accident
- Details of vehicle involved: Registration number, description of vehicle, loading detail, vehicular defects
- Nature of accident: Details of collision, damages, injury, and casualty
- Road and traffic condition: Details of road geometry, surface characteristics, type of traffic, traffic density, etc.
- Primary causes of accident: Details of various possible cases that are the main causes of accident.

These data collected need proper storing and retrieving for the purpose of identification of location of points at which unusually high number of accidents occur; detailed functional evaluation of critical accident location to identify the causes of accidents; development of procedure that allows identification of hazards before large number of accidents occur; and development of different statistical measures of various accident-related factors to give insight into general trends, common casual factors, driver profiles, etc.

BIBLIOGRAPHY

Global Status Report on Road Safety. (2023). World Health Organization.

Road Accidents in India. (2023). *Ministry of Road Transport and Highways (MoRTH).* Transport Research Wing, New Delhi.

Report on Road Environment Safety (1996): A Practitioners Reference Guide to Safer Roads, Road Traffic Authority of New South Wales, ISBN: 0731054423.

Guo, Y., Hussein, M., Chen, F., and Lyu, N.-C. (2021). "Surrogate safety measures in traffic safety analysis." *Journal of Advanced Transportation*, Special Issue 673933, 1–17, Wiley Publishers.

2 Road Safety Scenario and Influencing Factors

2.1 INTRODUCTION

Road accidents are one of the major reasons for loss of life or severe/minor injuries or property damage, and these are not constrained to a particular location on the globe. Human errors are one of the major causes of road accidents and these errors include over speeding, distractions to drivers, drunken driving, avoiding safety gears, and red light jumping. Not only human errors but other factors (weather conditions, road conditions, etc.) also influence the road accidents and a lot of research is going on to identify the factors that cause the road accidents.

India is the second largest country in the world in terms of both population and road network. With the growing number of vehicles and increase in road length, accidents resulting in injuries and fatalities have been mounting over the past 20 years. One serious road accident occurs every minute and one death occurs every four minutes due to a road accident in India. The Gross Domestic Product of any country is highly influenced by the road accidents and India loses 3% of its Gross Domestic Product only due to the road accidents. As per the statistics of the Ministry of Road Transport and Highways, total road accidents occurred, number of persons killed, and number of persons injured over the last five decades were shown in Figure 2.1. A drastic decrease in the total number of accidents and number of injuries, and almost equal number of road accident deaths every year from 2010 to 2020 was observed. This drastic change in the number of accidents and number of injuries is due to the improvements in the road traffic issues for smooth traffic flow. In 2020, a total of 3,66,138 accidents happened on Indian roads, which claimed 1,31,714 lives and caused 3,48,279 injuries. The number of persons killed in 2020 due to road accidents were 12.6% less than the number of persons killed in road accidents during 2019.

2.2 ROAD FACTORS

The geometry of road and pavement characteristics influence the occurrence of accidents. Tire-pavement interaction plays a key role in reducing the severity of accidents. The road surface characteristics like frictional properties, road condition, type of pavement, and condition of pavement have a high bearing on the cause of the accidents. A very smooth surface with less amount of friction must be avoided and at the same time a corrugated surface may not be an option considering the road user comfort and convenience.

DOI: 10.1201/9781003604761-2

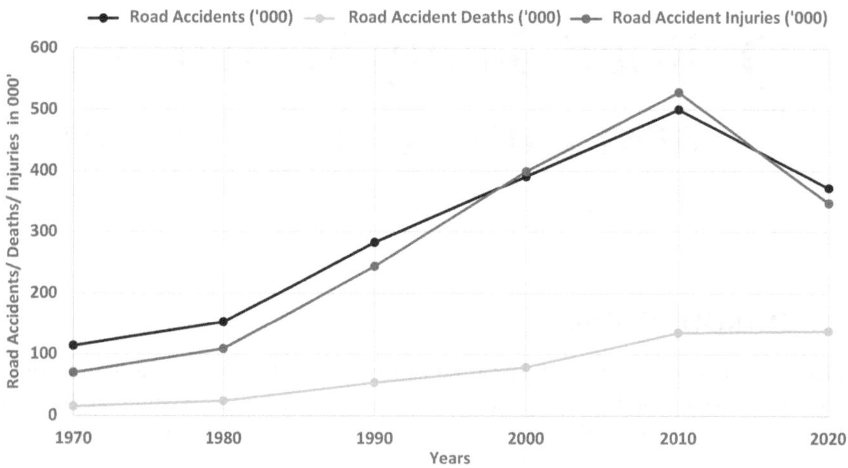

FIGURE 2.1 Decadal trend in the number of road accidents, deaths, and injuries from 1980 to 2020

Source: https://morth-roadsafety.nic.in/

The various road factors resulting in occurrence of accidents are listed as follows:

- Frictional characteristics of pavement surface
- Pavement condition and roughness
- Gradient of the road
- Carriageway elements like number of lanes, presence of shoulders and its width, and median presence.
- Road geometrical elements like sight distance, super elevation, and extra-widening provided.

The road design and geometrical elements play a crucial role in preventing accidents. A proper geometrical design catering to all types of road users will reduce the burden of road accidents. However, the design must be accompanied by an effective road planning exercise avoiding potential critical locations that are prone to possible blackspot locations. Sight distance plays a crucial role in preventing accidents and a proper sight distance triangle will have a crucial role in preventing the occurrence of road accidents. The intersection sight distance and proper line of sight visibility across the entire stretch will be crucial in having a proper road alignment.

2.3 VEHICLE FACTORS

Vehicular characteristics like braking efficiency, age of vehicle, vehicle condition, and regular maintenance influence the occurrence of accidents. The vehicular characteristics can be categorized into static and dynamic characteristics; static characteristics refer to vehicle length, breadth, height, turning radius, etc., whereas dynamic characteristics are speed, acceleration, engine power, etc.

The speed has a high bearing on the probability of an accident and is one of the major cause of most of the accidents and particularly crucial in the case of fatal accidents. A proper control of speed of vehicles is required from the road design and vehicular design point of view to reduce the effect of an accident in case of human error.

India's roads accommodate a wide variety of vehicles, including two-wheelers, cars, buses, trucks, bicycles, and even animal-drawn carts, along with pedestrians. This diversity leads to a chaotic environment at intersections. Nevertheless, issues such as poor lane discipline, improper roundabout use, and a lack of public awareness about the rules often result in confusion and accidents, highlighting the importance of context-specific studies and interventions. The rapid urbanization and growing vehicle densities across the world highlight the urgent need for effective traffic safety management solutions.

The vehicular design has grown widely in terms of safety perspective and in the recent years, number of assessment programs have been designed to ensure vehicles meet high safety standards in preventing accidents. Some of the recent developments in the safer vehicular technology are as follows:

- Bharat New Vehicle Safety Assessment Program: Crash tests cars to ensure safety compliance.
- Mandatory Safety Features: Airbags, anti-lock braking system, electronic stability control.
- Periodic Vehicle Fitness Tests: Older commercial vehicles must undergo fitness certification to remain operational.

2.4 ROAD USER FACTORS

The other major factor contributing to road accidents is road user factors. Road user characteristics such as vision, hearing, reaction time, and crossing speed play a major role in the driving activity as well as pedestrian activities. The reaction time of the driver is very much crucial in various tasks involved with vehicles. The reaction time may get influenced by the age, gender, experience, type of vehicle, and type of road. An inexperienced driver with little knowledge of road safety signs may require higher reaction times. Some of the recent studies observed that drivers in the young and middle-aged group are more likely to be involved in accidents. The reasons for this may be due to the higher involvement of this age group in the driving task and attitude levels. One more major cause for concern in the recent past is driver and pedestrian distraction due to various electronic devices.

Road user distraction refers to the behavior of road users using mobile devices, such as smartphones, for texting, talking, or listening to music while navigating on the road. This phenomenon poses significant challenges to road safety, as the use of electronic devices negatively impacts the attention and performance of road users. While much research has focused on the effects of such distractions on drivers, the growing prevalence of mobile phones has also raised concerns about their influence on pedestrians, particularly during critical tasks like walking and crossing streets.

TABLE 2.1
Major road user characteristics

Type	Characteristic	Recommendations
Visual	Horizontal view	0°–10° (Acute vision) 0–120° (Peripheral vision)
	Glare	Glare from opposite vehicles and recovery time takes 3 to 6 seconds
	Color	Ability to distinguish color variations
Hearing	Ability to hear sounds	No defined recommendations
Reaction time	Reaction of the driver	Useful in stopping the vehicle within safe distance and without colliding. Generally, 2.5 s for stopping sight distance and 2.0 s for overtaking operations.
Attention span	Ability to concentrate on the driving task	Average human attention span is 15 to 20 minutes and short-time lapses of looking away from the road for more than 2–3 seconds is major cause of crashes
Depth Perception	Ability to judge distances	Impaired ability to judge distances between vehicles or during overtaking operation

Digital distraction, such as using a mobile phone or looking at a digital billboard, can negatively impact driver safety at critical road sections.

Digital distractions impair situational awareness and reaction times, making both drivers and pedestrians more vulnerable to accidents. For pedestrians, activities like texting or listening to music can reduce their ability to detect traffic signals, hear approaching vehicles, or make quick, safe decisions at intersections. As mobile phone use continues to increase, addressing these risks has become a priority for improving overall road safety. Educational campaigns, technological solutions, and behavioral interventions are being explored to mitigate the impact of digital distractions on road users.

Drivers distracted by their phones tend to have reduced safety margins and exhibit wider pupil dilation compared to those who are focused on driving.

Pedestrians: Pedestrians distracted by digital devices exhibit slower reaction times when nearing crossing locations, intersections, and roundabouts.

The major road user characteristics and their recommended values are presented in Table 2.1.

2.5 TRAFFIC FACTORS, ENVIRONMENTAL FACTORS, AND OTHER FACTORS

Traffic density plays a crucial role in the occurrence of accidents. An expressway section or highway section with lesser traffic may result in higher vehicular speeds resulting in severe accidents. On urban road sections with denser traffic, speeds are generally lower; however, frequent stop-and-go operations in congested situations may test the road user patience levels resulting in accidents. Intersections are one of the major locations for the occurrence of accidents. Type of intersection plays a crucial role, and, in particular, uncontrolled intersections are more prone to accidents due to the lesser control of speed of vehicles. At signalized intersections, violation of red signal is one of the major causes of accidents. At roundabouts, non-confirmation of the rules of traversing around the roundabout may result in the occurrence of road crashes.

Environmental characteristics such as presence of fog and snow may result in low visibility conditions and result in fatal accidents in inclement weather. Presence of rain may result in slippery surface and loss of proper grip between tire and road surface. This may result in aqua planning effect resulting in fatal accidents.

Other factors like the presence of poles, billboards, and sign boards may obstruct the line of sight of road users and may result in crash occurrence.

2.6 COMBINATION OF FACTORS/MULTIPLE INFLUENCING FACTORS

Considering the role of various factors and their contributing cause of road accidents as described above, in reality most of the fatal accidents are due to multiple factors. For example, on a rainy day, a driver speeding at a higher speed above 60 km/h may result in lesser frictional coefficients between the road and tire surface resulting in loss of grip leading to a fatal accident. A proper reduction of speed might have reduced the severity of this accident. In most of the cases, a driver error or misjudgment combined with higher speeds and lack of proper signages may result in fatal accidents. For example, at an uncontrolled intersection, a car traveling at a very high speed nearly 100 km/h may collide with another vehicle from the other leg of the intersection due to a sight distance constraint. Considering these factors, there is a need to critically analyze the cause of accidents and the factors resulting from the various causes such as road, vehicle, and road user factors.

The accident cause-and-effect relation is very complex in nature and in some of the accidents; the real cause of the accident cannot be fathomed considering the gravity of the situation and likely tampering of accident scene due to the rescue operations. Nowadays, security surveillance cameras are available at major locations and offer a valuable information regarding the actual cause of accidents. However, on highway sections and other places, it is very difficult to ascertain the real cause of the accidents.

A proper accident reconstruction exercise from different perspectives needs to be planned to analyze an accident and suggest possible remedial measures. Nowadays, driver's distraction is becoming widely prevalent across the vehicle categories and one of the reasons is higher usage of mobile phones while driving and walking.

2.7 PERSPECTIVES

The perspective of the cause of a road accident must be looked at multiple angles considering the different factors discussed in the above sections. In most of the accidents, the blame game starts with the fault of road user, however, the real cause must be looked at from the perspective of all the stakeholders and considered in ascertaining the true cause of accident. One such case study is discussed below, which gives a proper perspective on deciding the real cause of an accident.

2.8 A REAL CASE STUDY OF AN ACCIDENT

A public transportation bus hit a passenger car at the T-junction on NH 163 resulting in three fatalities.

1. The bus was traveling at around 70–80 km/s just before the collision as per the GPS track provided by the authorities.
2. As per the information provided to the study team, the car was traveling from Hanamkonda to Hyderabad and they took a U-turn to visit the nearest Dhaba and came in the wrong direction to make a right turn at the Meedikonda T-intersection. While the car was trying to make a right turn, the bus collided with the car.
3. As per the information provided by the study team, the car appeared in the direction of path of bus resulting in fatal collision.
4. The point of contact is by the front of bus with the right side portion of the car.
5. After collision, the car was pushed onto the median and the bus leaned toward the opposite side of the highway section as shown in Figure 2.2.
6. It seems there was vegetation on the median section at the time of accident and the vegetation was trimmed subsequently at the time of study team visit.

Comments

1. As per the geometry of the highway section, the Meedikonda T-intersection is located at the end of the horizontal curve and there is possibility that the drivers on the highway section may have lesser visibility of vehicles waiting to make a turn at the median opening.

FIGURE 2.2 Image of the accident location

FIGURE 2.3 Image of the curved portion ahead of the accident location

2. There was a level difference in the two directional movements on the highway section and there is a high chance that the right turn from Warangal side on the highway section will have lesser sight distance of the oncoming traffic from Hyderabad leg of the intersection. This was observed during the field visit and there is a high possibility that the lesser sight distance will cause possible collision at this T-intersection.
3. The basic drawback with the geometry is that the T-intersection is combined with the horizontal curve and these kinds of situations always result in line of sight obstructions as shown in Figure 2.3.
4. There was a chance for misjudgment by the vehicle taking a right turn from the minor road due to lower conspicuousness of the vehicles on the horizontal curved section. The vehicles will cross half way and wait for the gap to merge with the traffic stream moving from Warangal to Hyderabad. This will result in obstructing the line of path of vehicles coming from Hyderabad to Warangal particularly on the median lane and resulting in collisions with right-turning vehicles.

Recommendations

1. The T-intersection median opening has to be completed closed due to the sight distance limitations and a U-turn is recommended for the convenience of vehicles from the minor road.
2. A speed limit sign board of 40 km/h is recommended at the starting of the horizontal curved section at a distance of 50 m.
3. The night time visibility has to be improved at the location.

2.9 ARE ROAD ACCIDENTS REALLY PREVENTABLE?

Considering the multitude of factors, the question to be raised is whether road accidents are preventable? The answer lies in multiple sectors in terms of proper planning and design of road sections, effective implementation of safety practices, enhanced vehicular safety standards, and effective road safety education and awareness campaigns. The onus lies on planners and engineers, road safety authorities, as well as common road users. Unless otherwise a massive exercise involving major stakeholders in the road sector combined with an extensive safety campaign, the burden of road accidents cannot be reduced in the near future.

The road safety has to be embedded into the design from the planning stage itself and a proper safety audit at an early stage may lead to a well-designed alignment and reduce the extra burden of accidents in the future. An extensive safety audit at various stages of planning, design, construction, and before opening will take care of any kind of anomalies on the road sections and will immensely benefit the road transport sector. The vehicular manufacturers must be able to research and develop safety embedded systems in the vehicular technology and make a motto to safeguard all the vehicular occupants in case of any eventuality. The licensing system needs to be enhanced to ensure a proper skill checking and periodic awareness of drivers of all categories of vehicles. Here, use of augmented and virtual reality may become a handy tool in creating awareness and determining the skills of various types of road users. Finally, the onus lies on the common road user for a responsible and safe behavior on the road. Despite the best efforts of the designers, vehicle manufacturers, authorities, and road construction agencies, unless the general road user is not complying with the rules of the road and show a defensive driving strategy, road accidents cannot be curtailed soon.

2.9.1 Summary

The causative factors for road accidents are multi-factorial and the challenging task is to analyze the effect of different categorical factors and address the root causes of the problem. The various stakeholders like general road users, vehicle manufacturers, road authorities, academicians, researchers, and traffic police play a crucial role in analyzing the accident scenarios and suggest suitable remedial measures.

BIBLIOGRAPHY

International Transport Forum/OECD. (2022). *The Safe System Approach in Action: An Experience-Based Guide to Enhanced Road Safety*. OECD Publishing, Paris, https://doi.org/10.1787/ad5d82f0-en.

Road Accidents in India. (2023). *Ministry of Road Transport and Highways (MoRTH)*. Transport Research Wing, New Delhi.

Tiwari, G., and Mohan, D. (2021). *Transport and Safety: Systems, Approaches, and Implementation.* Springer Transactions in Civil and Environmental Engineering, Springer, Singapore. https://doi.org/10.1007/978-981-16-1115-5

3 Methods of Accident Data Collection

3.1 TRADITIONAL METHODS

Prior to the advent of sophisticated digital and automated technologies, the acquisition of accident data predominantly depended on conventional methodologies. These methodologies, many of which are still in active application, have historically yielded essential insights into collision conditions, underlying causation elements, and the related human, vehicular, and environmental aspects. Conventional methods remain important due to regulatory requirements, their cost-effectiveness in resource-limited settings, and their role in preserving standardized historical datasets crucial for longitudinal and trend analysis. Despite facing criticism for underreporting and delays in data availability, these methods remain the principal official source of road crash statistics in numerous countries, including India. The following subsections elaborate on four primary conventional data sources in detail: police and official accident reports, hospital and emergency medical service (EMS) documentation, insurance records, and manual accident investigation reports.

3.1.1 Police and Official Crash Reports

Police crash reporting constitutes one of the most longstanding and well-acknowledged ways for collecting road traffic accident data. In most areas, statutes mandate that any collision resulting in injury, death, or significant property damage be reported to law enforcement. In India, Section 134 of the Motor Vehicles Act, 1988 requires drivers involved in serious accidents to promptly notify the nearest police station. These reports form the essential dataset for national road safety statistics, including those aggregated annually by the Ministry of Road Transport and Highways (MoRTH).

Content and Structure of Police Reports: Police crash reports are typically standardized in content and format, encompassing the following components:

1. **Administrative Information**: Date, time, and location of the crash; case reference number; and details of the reporting officer.
2. **Crash Description**: Narrative account of events preceding and following the crash, supplemented with sketch maps and, where available, photographic evidence.
3. **Involved Parties**: Details of drivers, passengers, and pedestrians; driving license particulars; and vehicle registration information.

DOI: 10.1201/9781003604761-3

4. **Injury Details**: Classification of injury severity (fatal, grievous injury, minor injury, or non-injury).
5. **Contributory Factors**: Preliminary identification of potential causes, including driver behavior, vehicle mechanical defects, road surface conditions, and prevailing weather.

Police-reported road accident statistics constitute the legally recognized and authoritative record for legal procedures, insurance claims, and statistical analysis. It includes all documented incidents that satisfy the legal criteria for obligatory reporting. Furthermore, it employs the mandated formats to guarantee consistency in data collection and to enable comparative studies. Nonetheless, police-reported data has numerous drawbacks, particularly with non-fatal and minor crashes, which frequently go unreported to authorities (World Health Organisation, 2018). The variety in data quality, including the level of detail, may vary based on the training, experience, and workload of the reporting officer. Furthermore, delays may arise between the incidence of a crash and the submission of information into centralized systems, so constraining timely analysis.

3.1.2 Hospital and EMS Records

Medical data collected by hospitals, trauma centers, and Emergency Medical Service providers provides an essential and supplementary viewpoint to police crash investigations. These sources provide a comprehensive account of the clinical outcomes of road crash victims, frequently offering more extensive medical details than law enforcement records. Hospital admission records, emergency department logs, and ambulance service databases collectively offer significant insights on injury characteristics, treatment modalities, and patient outcomes.

Content of Medical Records: Hospital and EMS datasets typically include the following elements:

1. **Patient Demographics**: Age, gender, occupation, and place of residence.
2. **Clinical Information**: Nature, anatomical location, and severity of injuries; treatment provided; and final outcome (e.g., recovery, disability, or death).
3. **Mechanism of Injury**: Coded according to standardized systems such as the International Classification of Diseases *(ICD-10) injury codes.*
4. **Timestamps**: Times of arrival at the facility, initiation of treatment, and discharge or death.

Typically, hospital and EMS data exhibit high sensitivity for injury instances involving patients who did not notify law enforcement about the incident. Medical precision enables severity assessment by indices like the Abbreviated Injury Scale or the Injury Severity Score. The significance of public health underpins the necessity for burden-of-injury research and guides healthcare budget allocation. Hospital and EMS records data exhibit various shortcomings, including the absence of collision context, which frequently excludes facts about the crash scene, vehicle type, and environmental conditions. This data presents linkage issues, as connecting medical information with police records necessitates unique identifiers that are often inaccessible. Additionally, omit persons who do not pursue medical care.

3.1.3 Insurance Records and Claims Data

Insurance firms preserve comprehensive records of vehicle injury claims, personal injury claims, and the corresponding financial settlements. Although these databases are mostly intended for commercial and legal applications, they serve as a significant secondary source of information regarding crash frequency, severity, and economic repercussions.

Content of Insurance Records: Insurance datasets commonly include:

1. **Claimant and Policy Details**: Driver information, vehicle particulars, and policy coverage details.
2. **Accident Particulars**: Date, time, location, and type of collision.
3. **Damage Assessment**: Estimated repair costs, identification of total loss cases, and valuation of property damage.
4. **Medical Expenses**: Hospital bills, rehabilitation costs, and other health-related expenses (for injury-related claims).

Insurance records offer an economic perspective that provides immediate assessments of property damage and healthcare expenditures. Additionally, it encompasses extensive databases, particularly in nations with elevated automobile insurance penetration rates. Moreover, the possibility for fraud detection of insurance investigations may reveal orchestrated or deceptive collisions. Numerous constraints of insurance data omit uninsured automobiles and instances where repairs are financed privately. Furthermore, access to data is frequently limited by confidentiality and data protection requirements, so restricting privacy. Furthermore, discrepancies in non-standard formats of record-keeping among insurers impede aggregation and analysis.

3.1.4 Manual Crash Investigation (On-site Inspections and Reconstruction)

Manual crash investigation entails the direct, on-site analysis of accident sites performed by qualified professionals—such as law enforcement investigators, road safety engineers, or specialized forensic teams—aimed at reconstructing the sequence of events and identifying contributing variables.

3.1.4.1 Key Activities in Manual Crash Investigation

1. **On-site Inspections**: Measuring skid marks, yaw marks, debris distribution, and points of impact.
2. **Photography and Video Documentation**: Capturing vehicle positions, roadway features, and prevailing environmental conditions.
3. **Vehicle Examination**: Identifying mechanical defects or analyzing post-impact damage patterns.

4. **Witness Interviews**: Collecting firsthand accounts to complement physical evidence.
5. **Reconstruction Analysis**: Applying physics-based methods to estimate vehicle speeds, trajectories, and collision forces.

Manual crash investigations provide significant contextual details that encompass important elements not evident from administrative records alone. It also possesses legal robustness, offering admissible proof for legal and insurance processes. Furthermore, this data typically encompasses engineering insights that identify hazardous road designs or maintenance issues, facilitating targeted interventions. The primary drawbacks of manual crash investigations include their resource-intensive nature, the necessity for qualified personnel, specialized equipment, and considerable time commitment. Potential bias stemming from investigator interpretation may affect conclusions. Furthermore, practical limitations on the quality of evidence may be undermined if inspections are postponed.

3.2 ADVANCED METHODS

Recent developments in information technology, digital communication networks, and data analytics have fundamentally altered the framework of road traffic accident data collection. Although conventional paper-based and manual reporting methods remain relevant in some circumstances, they are often impeded by delayed reporting, format discrepancies, and incomplete coverage. Contemporary, technology-driven approaches mitigate these restrictions through the use of digital platforms, automated data collection, and cohesive databases. These improvements enhance the efficiency of documenting collision events while also augmenting data accuracy, accessibility, and timeliness. The advanced collection approaches facilitate the creation of more comprehensive and dependable datasets, thereby enhancing the foundation for evidence-based decision-making in road safety management. They also promote inter-agency coordination, bolster advanced analytics, and improve stakeholders' capacity to devise and execute targeted actions.

3.2.1 Electronic Data Systems (e-CRFs and Digital Submission Platforms)

Electronic data capture solutions, encompassing electronic case report forms (e-CRFs) and specialized digital reporting portals, are increasingly supplanting traditional paper-based crash documentation. These platforms enable authorized personnel—such as law enforcement officials, emergency responders, and transportation department staff—to document incident facts directly into organized computer templates, often in real time.

3.2.1.1 Key Features of Electronic Data Systems

1. **Standardized Templates**: Predefined forms ensure uniformity in data collection across jurisdictions.
2. **Real-Time Data Transfer**: Enables immediate transmission of crash information to central databases, facilitating rapid analysis.

3. **Built-in Validation Checks**: Mandatory fields, dropdown selections, and logical constraints (e.g., prohibiting crash dates set in the future) help reduce data entry errors.
4. **Global Positioning System (GPS) and Time-Stamping**: Provides precise geographic and temporal information for each reported event.
5. **Multimedia Integration**: Allows photographs, videos, and diagrammatic sketches to be attached directly to the digital report.

The rapid transfer of information facilitates prompt monitoring of crash trends. Typically, improved precision that eradicates transcription errors associated with the conversion of paper records into digital versions. Moreover, enhanced accessibility that ensures secure, remote access for authorized parties promotes inter-agency coordination.

3.2.2 Integrated Databases and Data Linkage Techniques

Integrated accident databases amalgamate data from several sources, including police crash reports, hospital admission records, insurance claims, and traffic flow statistics, into a unified dataset. This approach frequently depends on data linkage methodologies that correlate associated records across agencies utilizing identifiers such as crash IDs, vehicle registration numbers, or anonymized patient codes.

3.2.2.1 Importance in Road Safety Research and Management

Single-source datasets are often incomplete. For example:

- Police data may detail the circumstances of a crash but omit medical outcomes.
- Hospital data may provide clinical details yet lack information on crash causation.
- Insurance data may record property damage and associated costs but not injury severity.

By integrating these datasets, researchers and policymakers may develop a more thorough and precise profile of each collision incident. The primary advantages facilitate comprehensive epidemiological investigations of injury patterns and consequences. It enhances precision in severity classification through the integration of medical data. Additionally, it enables more accurate cost evaluations by integrating property damage and healthcare expenditure data.

3.2.3 Statistical and AI-Based Data Cleaning and Underreporting Adjustments

Even with advanced digital reporting tools, accident datasets may contain missing values, inconsistencies, or biases caused by underreporting. Minor crashes, rural incidents, and cases where stakeholders lack incentive to report are particularly prone to omission. Statistical and artificial intelligence (AI) techniques offer solutions for enhancing data quality and estimating the true scale of crash incidence.

3.2.3.1 AI Applications in Data Cleaning

- **Natural Language Processing**: Converts unstructured narrative descriptions from crash reports into structured, analyzable variables.
- **Machine Learning-Based Error Detection**: Identifies implausible records, such as improbable vehicle speeds or mismatched geographic coordinates.
- **Automated Geocoding**: Corrects and standardizes crash location data using GPS and digital mapping services.

3.2.3.2 Methods for Underreporting Adjustment

1. **Capture–Recapture Analysis**: Originally developed for ecological studies, this method estimates the total number of crash events by comparing overlaps between independent datasets (e.g., police vs. hospital records).
2. **Bayesian Hierarchical Models**: Incorporate prior knowledge about reporting rates to adjust observed crash counts.
3. **Imputation Techniques**: Use statistical models trained on complete cases to estimate missing data, such as injury severity.

AI-based data produces more accurate crash incidence estimates for planning and resource allocation. Generally, it improves comparability of statistics across jurisdictions and time periods. Also, it supports predictive modeling, enabling proactive safety interventions.

3.3 DRONE APPLICATIONS

The advent of unmanned aerial vehicles (UAVs), usually known as drones, has considerably broadened the resources accessible for transportation engineering and road safety management. Drones offer an airborne perspective for accident data collection that was previously restricted to expensive helicopter operations. This technical innovation facilitates swift, precise, and economical documentation of accident scenes, aiding both reactive functions (post-accident investigation) and preventative applications (risk evaluation at perilous locations). Drone-based observation reconciles traditional ground-based data collection, which is frequently labor-intensive and prone to observer bias, with stationary automated sensing systems that necessitate significant infrastructure investment. UAV platforms provide adaptability, high-resolution photography, and operational efficacy in situations where manual data collection is either hazardous or logistically difficult.

3.3.1 Key Advantages of Drone Technology

a. **Rapid Deployment and Spatial Coverage**
   Drones can be deployed within minutes of a crash notification, enabling investigators to document the site prior to alterations caused by emergency response efforts, rescue operations, or traffic clearance. The overhead

perspective offers a detailed spatial picture, encompassing adjacent roadway geometry, traffic control systems, and pertinent environmental characteristics inside a unified dataset.

b. High-Resolution Imaging and Mapping
Modern UAVs are outfitted with high-definition camera equipment, frequently capable of 4K video recording and thermal photography. These sensors record intricate data including skid marks, impact damage on cars, and patterns of debris dispersion. When integrated with photogrammetry and Structure-from-Motion processing, drone imagery can provide accurate 3D reconstructions of crash sites for forensic examination.

c. Safety and Accessibility
The utilization of drones diminishes the necessity for investigators to physically access perilous environments, like high-velocity traffic zones, unstable landscapes, or regions impacted by hazardous material leaks. UAVs can reach remote or obstructed crash sites, such as rural roadways, mountainous regions, and flood-affected routes, which may be unreachable to traditional survey teams.

d. Data Integration Capabilities
Drones can collect imagery and spatial data that can be georeferenced and integrated into Geographic Information Systems (GIS), electronic crash reporting forms, or centralized police systems. This integration creates a cohesive dataset that enhances multi-agency collaboration and allows for the dissemination of standardized accident reports for legal, administrative, and research applications.

3.3.2 Applications in Accident Scene Investigation

a. Scene Documentation
UAVs offer comprehensive, aerial, and oblique photography, documenting the whole accident site within its spatial context. Aerial imaging reduces perspective distortion and improves the precision of spatial relationship interpretation compared to conventional ground-level photography. This facilitates the precise identification of the definitive final locations of vehicles, debris fields, impact points, and roadside hazards or impediments affecting crash dynamics.

b. Evidence Preservation
Accident scenes can evolve swiftly due to meteorological conditions, continuous vehicular movement, or emergency responses. Drones maintain the site's original condition by capturing high-resolution video and still images, so establishing a permanent digital record appropriate for legal procedures, insurance assessments, and scholarly research.

c. **Measurement and Reconstruction**
Utilizing orthomosaic imagery and, where accessible, Light Detection and Ranging (LiDAR)-equipped UAVs, researchers can acquire accurate measurements of the length and orientation of skid or yaw marks, lane widths, shoulder conditions, and roadway characteristics, as well as the distances between impact points and final vehicle rest positions. These observations augment collision reconstruction models, hence enhancing the precision of speed estimation and trajectory analysis.

d. **Multi-Temporal Analysis**
UAVs can be used in accident-prone sites over several timeframes to evaluate temporal fluctuations in risk factors, like alterations in illumination, traffic density, or roadside foliage. This method facilitates the assessment of seasonal or diurnal effects on crash probability.

3.3.3 Applications in Proactive Road Safety Management

Drones are being utilized as a preventive safety measure, identifying risks prior to their contribution to accidents, in addition to their function in post-crash examinations. Principal proactive applications encompass:

- **Work Zone Monitoring**: Frequent aerial surveillance of road construction and maintenance areas to verify adherence to safety regulations and detect hazardous conditions in real time.
- **Black Spot Identification**: Regular aerial reconnaissance to identify areas with persistent high-risk characteristics, including limited visibility, deteriorated or insufficient signage, and unlawful roadside parking.
- **Traffic Flow Assessment**: UAV-based video collection facilitates the examination of traffic stream attributes, encompassing speed distribution, headway patterns, and hazardous driving behaviors such as sudden lane changes or tailgating.

3.4 INTEGRATED ROAD ACCIDENT DATABASE

The Integrated Road Accident Database (iRAD) is India's initial comprehensive, technology-driven, multi-agency platform for managing accident data. Developed by the Indian Institute of Technology Madras in conjunction with the National Informatics Centre and initiated by the MoRTH, the system signifies a transformative change in the nation's methodology for collecting, integrating, and utilizing road crash data. Supported by the World Bank under its extensive road safety capacity building programs, iRAD seeks to rectify persistent inadequacies in the quality, timeliness, and accessibility of accident data. Before the implementation of iRAD, accident data in India was mostly derived from police First Information Reports and Accident Information Reports, which frequently experienced underreporting, inconsistency, and delays in aggregation. The disjointed distribution of data among police, transport, health, and road management authorities impeded prompt decision-making.

iRAD has been developed to address these limitations by establishing a cohesive, digital, and geospatially enabled database that consolidates contributions from all pertinent parties.

3.4.1 iRAD Objectives

The establishment of iRAD is guided by the following primary objectives:

1. **Holistic Data Integration**: To consolidate accident-related information from police, transport, health, and road-owning agencies into a single platform.
2. **Improved Data Quality**: To ensure completeness, accuracy, and reliability of accident data by incorporating geotagging, multimedia evidence, and standardized input formats.
3. **Support for Evidence-Based Policy**: To enable state and national authorities to identify crash trends, high-risk locations, and contributory factors for targeted interventions.
4. **Facilitation of Legal and Insurance Processes**: To streamline accident claim settlements and medical reimbursement procedures through timely availability of verified data.
5. **International Benchmarking**: To align India's crash data collection practices with global systems such as the US Fatality Analysis Reporting System and the European CARE database. (Care is a community database on road accidents resulting in death or injury).

3.4.2 System Architecture and Workflow

The iRAD platform has been designed as a modular, interoperable, and scalable system that can function in diverse state environments. Its architecture includes the following components:

a. Data Capture Layer

Accident details are initially recorded by field personnel—primarily police officers—using the iRAD mobile application. The application allows:

- Automatic geotagging of accident locations.
- Uploading of photographs and videos from the scene.
- Capture of preliminary crash details such as date, time, weather, and road conditions.

Once the initial record is created, a unique Accident ID is generated, enabling linkage of supplementary information from other agencies.

b. Multi-Agency Data Contribution

Following the initial police entry:

- **Transport Departments** provide vehicle and driver details, drawing from VAHAN and SARATHI databases.
- **Health Departments** input injury severity, treatment details, and outcomes, facilitating linkage to trauma care data.

- **Road-Owning Agencies** (National Highways Authority of India, State Public Works Department (PWDs), Municipal Corporations) contribute road geometry, maintenance history, and traffic control device information.

c. **Data Validation Layer**
Each agency verifies its respective data entries before final submission. The platform's built-in validation checks help ensure consistency, reduce typographical errors, and flag incomplete fields.

d. **Data Analysis and Visualization**
A web-based **Road Safety Dashboard** offers:

- Real-time accident mapping via GIS.
- Time-series trend analysis for crash frequency, severity, and location.
- Identification of high-risk corridors and black spots.
- Automated generation of periodic statistical reports for state and national use.

3.4.3 Implementation Strategy

The national rollout of iRAD followed a phased approach:

1. **Pilot Phase**: Initiated in select districts across six states to test functionality, gather user feedback, and refine the application interface.
2. **Capacity Building**: Training of Trainers workshops conducted for police, transport, health, and road authority officials.
3. **Statewide Expansion**: Gradual onboarding of districts until full state coverage was achieved.
4. **Integration with Existing Systems**: Establishment of data exchange protocols with VAHAN, SARATHI, and MoRTH's National Accident Database.

Comprehensive coverage encompasses all accidents recorded by various authorities, thereby mitigating underreporting. The data will encompass multimedia, geospatial coordinates, and corroborated agency inputs. Furthermore, timeliness facilitates near-instantaneous data entry and retrieval. iRAD data promotes a collaborative methodology for road safety management and equips policymakers with practical insights for infrastructure development, enforcement tactics, and public awareness initiatives. Nonetheless, numerous issues in iRAD encompass the underreporting of small accidents, akin to conventional police reporting systems, where many non-fatal collisions remain undocumented. In specific jurisdictions, insufficient personnel impede prompt data entry. Furthermore, disparities in digital competencies among field personnel can influence data integrity. In particular, in rural and remote regions, inadequate network coverage can hinder uploads, while a deficient workforce leads to more errors. Furthermore, addressing data privacy concerns is crucial as the system amalgamates personal, medical, and legal information, necessitating adherence to privacy regulations.

3.4.4 Summary

Collecting accident data is an essential measure for improving road safety across the world. In the absence of reliable data, it is impracticable to comprehend the mechanisms, locations, and causes of accidents, or to devise engineering solutions to prevent them. Dependable data allows authorities to identify perilous road locations, discern common accident causes, and assess progress after the introduction of safety interventions. Worldwide, vehicle accidents cause more than 1.35 million fatalities each year (WHO, 2018), with tens of millions more suffering injuries. In India, more than 172,000 deaths occur annually as a result of traffic accidents (MoRTH, 2023). The significant social and economic burden necessitates the immediate prioritization of collecting accurate, timely, and comprehensive crash data. Historically, the collecting of accident data depended on manual, paper-based methods: law enforcement documented crash scenes, medical facilities recorded injuries, and insurance companies processed claims. Although traditional methods remain significant—particularly in regions with inadequate digital infrastructure—they are progressively augmented or substituted by sophisticated technologies such as automated sensors, CCTV analytics, telematics, drones, and integrated databases.

BIBLIOGRAPHY

e-Challan - Digital Traffic/Transport, Enforcement Solution, (2025, November 14), https://echallan.parivahan.gov.in/echallanreport//

Integrated Road Accident Database, (2025, November 14), https://irad.parivahan.gov.in/

MoRTH, 2022. Road Accidents in India-2022, Ministry of Road Transport and Highways (MoRTH), Transport Research Wing, India. https://morth.nic.in/sites/default/files/RA_2022_30_Oct.pdf

Ministry of Road Transport & Highways, Government of India, (2025, November 14), https://vahan.parivahan.gov.in/vahan/vahan/ui/login/login.xhtml

WHO, 2018. Global status report on road safety 2018. Geneva: World Health Organization.

Zhu Meng, Ning Zhu, Guowei Zhang, Yuance Yang, Zhaocai Liu, and Ginger Y. Ke. (2025, November, 14). "Data-driven drone pre-positioning for traffic accident rapid assessment." *Transportation Research Part E: Logistics and Transportation Review*, 183(2024), 103452, ISSN 1366–5545. https://doi.org/10.1016/j.tre.2024.103452

4 Methods of Accident Analysis

4.1 INTRODUCTION

The **methods of accident analysis** can be broadly classified into:

1. **Statistical Methods**: including descriptive and inferential statistics to identify crash patterns, trends, and risk factors.
2. **Advanced Analytical Techniques**: such as Bayesian modeling, survival analysis, and machine learning (ML), which allow for more robust predictive modeling.
3. **Conflict Analysis**: focusing on near-miss situations and surrogate safety measures (SSMs) to anticipate crashes before they occur.
4. **Data Fusion and Exploration**: integrating multiple data sources [e.g., crash data, traffic volumes, weather, and Global Positioning System (GPS) traces] to provide a holistic understanding of safety risks.

4.2 STATISTICAL ANALYSIS

Statistical analysis forms the foundational core of modern accident research, serving as the crucial bridge between raw, often unwieldy, crash data and the actionable knowledge required for effective road safety policy and engineering. The inherent complexity of crash data—characterized by its large volume, heterogeneity, and the multitude of factors at play—necessitates a rigorous statistical approach to extract meaningful insights. In the context of road safety, this data typically encompasses a diverse range of variables, including:

- **Crash Frequency**: The numerical count of crashes occurring at a specific site (e.g., an intersection), over a defined time period (e.g., per year).
- **Crash Severity**: The classification of crash outcomes, typically ranging from fatal or serious injury to minor injury and property damage only.
- **Exposure Measures**: Variables that account for the volume of activity, such as vehicle-kilometers traveled or average daily traffic, which are essential for normalizing crash counts to represent risk.
- **Contributing Risk Factors**: A broad spectrum of variables that may influence crash likelihood or severity, including road geometry (e.g., curvature, grade), environmental conditions (e.g., weather, lighting), driver demographics (e.g., age, experience), and vehicle characteristics.

DOI: 10.1201/9781003604761-4

4.2.1 Basic Statistical Concepts Applied in Accident Analysis

- **Descriptive Statistics**: Measures like **mean, median, mode, standard deviation, skewness, and kurtosis** summarize crash characteristics. For instance, computing the mean crash frequency across sites helps identify outliers such as "black spots."
- **Inferential Statistics**: Hypothesis testing, chi-square tests, and analysis of variance (ANOVA) are used to compare safety performance across locations or time periods.
- **Regression Models**: Techniques such as Poisson regression or Negative Binomial (NB) regression are widely applied to model crash counts while accounting for exposure.
- **Probability Distributions**: Crash occurrences often follow discrete probability distributions (e.g., Poisson for rare events, NB for over-dispersed crash counts).

4.2.2 Descriptive Statistics

Descriptive statistics are the foundation of accident data analysis, as they provide an initial summary of the crash dataset. Instead of attempting to establish causal relationships at this stage, descriptive methods organize, simplify, and present raw accident data in a meaningful form. This enables researchers, policymakers, and engineers to recognize patterns such as accident frequency, severity trends, high-risk locations, or common vehicle types involved.

4.2.2.1 Key Measures in Descriptive Statistics

4.2.2.1.1 Measures of Central Tendency

Central tendency identifies the "average" or most representative value in a dataset of accidents (e.g., number of crashes per day, average age of drivers in fatal crashes).

Mean:

The mean is the sum of all data values divided by the number of observations.

$$\text{Mean } (\bar{x}) = (\textstyle\sum xi) / n \tag{4.1}$$

Example: Suppose we record the number of crashes at an intersection over 7 days: $x = \{5, 7, 6, 8, 10, 4, 5\}$

Mean = (5 + 7 + 6 + 8 + 10 + 4 + 5)/7 = **6.43 crashes/day**

Median (Middle Value):

The median is the middle value when data are arranged in ascending order.

- If ***n*** **is odd**: Median = middle observation.
- If ***n*** **is even**: Median = average of two middle observations.
- **Example**: Accident counts per day: {4, 5, 5, 6, 7, 8, 10}

Median = 6 crashes/day.

Mode (Most Frequent Value):

The mode is the value that occurs most frequently in the dataset.

- **Example**: In the dataset {5, 7, 6, 8, 10, 4, 5},
  Mode = 5 (since it repeats twice).

4.2.2.1.2 Measures of Dispersion (Spread of Data)

Dispersion shows how widely accident data values vary around the central tendency.

Range:

Range in a statistical context is a measure of dispersion or spread. It's a very simple concept: it's the difference between the highest and lowest values in a dataset.

$$\text{Range} = \text{Maximum value} - \text{Minimum value}$$

- **Example:** For daily crashes {4, 5, 5, 6, 7, 8, 10},
  Range = 10 – 4 = **6**

Variance (σ^2):

Variance represents the average of squared deviations from the mean.

$$\sigma^2 = \frac{\sum_{i=1}^{n} (x_i - \bar{x})^2}{n} \tag{4.2}$$

Standard Deviation (σ):

Standard deviation is the square root of variance, providing spread in the same units as the data.

$$\sigma = \sqrt{\sigma^2} \tag{4.3}$$

Example: For the given data {5, 7, 6, 8, 10, 4, 5} estimate the mean, variance, and standard deviation of accidents?

Mean = 6.43

Variance = 3.81

Standard Deviation = 1.95

Interpretation: Accident counts vary about **±2 crashes/day** around the average.

4.2.2.1.3 Measures of Shape

These describe the shape of accident data distribution.

a. Skewness: Skewness measures asymmetry of the data.

$$\text{Skewness} = \frac{\sum (x_i - \bar{x})^3}{n\sigma^3} \tag{4.4}$$

- Positive skew → long right tail (e.g., accident costs where few severe crashes inflate values).
- Negative skew → long left tail.

b. Kurtosis:

Kurtosis indicates how peaked or flat the distribution is compared to a normal distribution.

$$\text{Kurtosis} = \frac{\sum (x_i - \bar{x})^4}{n\sigma^4} \tag{4.5}$$

- High kurtosis → more outliers, sharper peak.
- Low kurtosis → flatter distribution.

4.2.3 Frequency Distributions and Cross-Tabulations

Accident data, when gathered in its unprocessed state, generally presents as a vast and disorganized compilation of incidents. Researchers and policymakers find it challenging to directly interpret such data. Fundamental statistical tools like frequency distribution and cross-tabulation assist in structuring this information. These tools enable analysts to uncover patterns, pinpoint high-risk groups, and gain insights into accident features such as severity, location, timing, and contributing factors. While frequency distributions provide a summary of how frequently a certain category or range occurs, cross-tabulations take this a step further by illustrating the interactions between two or more variables. Together, these approaches lay the groundwork for descriptive analysis of accident data.

4.2.3.1 Frequency Distributions

Frequency distributions are one of the most widely used descriptive techniques in accident data analysis. They provide a simple yet powerful way to summarize large datasets by showing how often accidents occur within specific categories (e.g., by time, location, severity). By transforming raw accident records into structured summaries, researchers can detect patterns such as peak accident hours, high-risk vehicle categories, or road surface conditions associated with crashes.

For example, a frequency distribution can be applied to answer essential questions such as:

- What is the count of crashes distributed across different roadway classifications (e.g., freeway, arterial, or local street)?
- How is crash severity distributed across the dataset (e.g., fatal, serious injury, minor injury)?
- Which time periods (e.g., time of day or day of the week) exhibit the highest frequency of accidents?

4.2.3.2 Mathematical Representation

A frequency distribution displays the number of observations falling into specific categories or class intervals. If n represents the total number of observations, the **absolute frequency** f_j of class j is given as:

$$\sum_{j=1}^{k} f_j = n \tag{4.6}$$

Where:
f_j = frequency of class j
k = number of classes

The **relative frequency** is defined as:

$$r_j = \frac{f_j}{n} \tag{4.7}$$

The **percentage frequency** is given by:

$$\%f_j = \frac{f_j}{n} \times 100 \tag{4.8}$$

The **cumulative frequency** is:

$$CF_j = \sum_{i=1}^{j} f_i \tag{4.9}$$

These measures enable both absolute and proportional understanding of accident occurrences.

Steps in Constructing Frequency Distributions

1. **Define the variable of interest** (e.g., accident severity, time of day, vehicle type).
2. **Group data into classes or categories** (for numerical variables, select class intervals such as 0–10, 10–20).
3. **Count the number of observations** in each class (absolute frequency).
4. **Compute relative and cumulative frequencies** for better interpretation.
5. **Present results** using a frequency table or a histogram.

4.2.3.3 Cross-Tabulation

A cross-tabulation known as contingency table is a fundamental analytical tool for exploring the relationship between two or more categorical variables. This powerful method arranges data in a matrix, with the categories of one variable forming the rows and those of another forming the columns. Each cell at the intersection of a row and a column contains the frequency of observations that share both categories.

4.2.4 Time-Series Analysis

Time-series analysis is a statistical technique used to analyze accident data observed over time. Road accident frequencies, severities, or rates often exhibit temporal patterns such as trends, seasonality, and cycles. Understanding these components is essential for predicting future accident risks and for designing targeted safety interventions.

Components of Time-Series in Accident Data:

1. **Trend (T)**: Long-term increase or decrease in accident frequency. Example: A steady decline in road fatalities due to stricter enforcement.
2. **Seasonality (S)**: Repeated patterns within a fixed period, e.g., accidents peaking during monsoon months or holiday seasons.
3. **Cyclic Variation (C)**: Longer-term oscillations, often linked to economic activity, fuel prices, or vehicle ownership.
4. **Irregular Component (I)**: Random fluctuations caused by unforeseen factors such as sudden weather changes or natural disasters.

4.2.5 General Model

$$Y_t = T_t + S_t + C_t + \epsilon_t \tag{4.10}$$

Where:

- Y_t = observed accident frequency at time t
- T_t = long-term trend
- S_t = seasonal component
- C_t = cyclic variation
- ϵ_t = random error

where YtY_tYt = observed accidents at time ttt.

Methods Used

- **Moving Averages**: Smooths short-term fluctuations to identify trend.
- **Exponential Smoothing (Holt–Winters)**: Gives higher weight to recent observations.
- **Auto-Regressive Integrated Moving Average**: Widely applied for accident forecasting.

4.2.6 Regression Models

Regression analysis is a widely applied statistical technique in road safety research. Its purpose is to establish and quantify the relationship between the number of accidents (the dependent variable) and various explanatory variables (independent variables) such as traffic flow, speed, road geometry, weather, pavement condition, and roadside environment. By identifying how these factors influence crash occurrence, regression models can be used both for understanding accident causation and for predicting future accident frequencies under different roadway and traffic conditions. The key challenge in accident prediction is that crash data are non-negative, discrete, and highly variable. Therefore, the choice of regression model must be carefully aligned with the statistical nature of the data.

4.2.6.1 Linear Regression

Linear regression is the simplest regression model and has historically been used in accident prediction. In this approach, accident frequency is treated as a continuous variable, even though in reality, crashes are non-negative counts.

The model is expressed as:

$$Y = \beta_0 + \beta_1 X + \in \tag{4.11}$$

Where:

- Y = number of accidents (dependent variable)
- X = explanatory variable [e.g., traffic volume or average annual daily traffic (AADT)]
- β_0, β_1 = model coefficients
- ϵ = random error term

4.2.6.1.1 Interpretation

- β_0: baseline number of accidents when X = Traffic volume in AADT
- β_1: change in expected number of accidents per unit increase in traffic volume.

For example, if $\beta_1 = 0.02$, then for every 1,000 vehicles/day increase in AADT, the model predicts 20 additional accidents per year.

4.2.6.1.2 Limitation

- **Negative Predictions**: Since the model assumes a continuous dependent variable, it may predict negative accident counts, which is unrealistic.
- **Ignoring Count Nature**: Accident data are non-negative integers (0, 1, 2,…), making linear regression statistically inappropriate.

- **Non-Constant Variance**: Accident data often show over-dispersion (variance > mean), violating the assumption of homoscedasticity in linear regression.

Despite its limitations, linear regression is still useful for preliminary analysis and as a baseline comparison against more advanced models.

4.2.6.2 Poisson Regression

Poisson regression is a statistical method used to model count data, which represents the number of occurrences of an event. It's a type of generalized linear model that assumes the response variable follows a Poisson distribution, characterized by a mean that equals its variance. This makes it suitable for analyzing data where the dependent variable is a count, such as the number of customer inquiries, website visits, or accidents. The probability distribution of accident counts is given by:

$$P(Y = y) = \frac{e^{-\lambda}\lambda^{y}}{y!} \tag{4.12}$$

Where:

λ = expected number of accidents (both mean and variance)
y = observed number of accidents

The expected accident frequency is modeled as:

$$\lambda = e^{\beta_0 + \beta_1 X_1 + \beta_2 X_2 + \ldots} \tag{4.13}$$

This ensures that λ is always positive, avoiding negative predictions.

4.2.6.2.1 Key Property

$$E(Y) = Var(Y) = \lambda \tag{4.14}$$

This means the mean and variance of crash counts are assumed to be equal.

4.2.6.2.2 Limitation

If accident data exhibit over-dispersion (variance > mean), the Poisson model underestimates variability, leading to biased results.

4.2.6.3 Negative Binomial Regression

In real-world accident data, variance is often larger than the mean due to unobserved heterogeneity (factors not captured in the model such as driver behavior or enforcement level). This condition is called over-dispersion, and in such cases, Poisson regression becomes inadequate. NB regression addresses this issue by adding a dispersion parameter (α\alpha) that allows the variance to exceed the mean:

$$\text{Var}(Y) = \mu + \alpha\mu^2 \tag{4.15}$$

The probability distribution is:

$$P(Y = y) = \frac{\Gamma\left(y+\alpha^{-1}\right)}{\Gamma(y+1)\Gamma\left(\alpha^{-1}\right)}\left(\frac{1}{1+\alpha\mu}\right)^{\alpha^{-1}}\left(\frac{\alpha\mu}{1+\alpha\mu}\right)^{y} \tag{4.16}$$

Where:

Y = crash frequency

μ = expected number of accidents

α = dispersion parameter (when $\alpha = 0$, the NB model reduces to Poisson).

4.2.7 Hypothesis Testing in Accident Analysis

The general procedure of hypothesis testing involves formulating a null hypothesis (H_0)—which assumes no effect or no difference—and an alternative hypothesis (H_1)—which represents the expected change or relationship. A test statistic is then calculated from sample data and compared against a critical value (or evaluated via a p-value) to determine whether the null hypothesis should be rejected.

Several commonly used hypothesis tests in accident analysis are described below.

4.2.7.1 *t*-Test

The ***t*-test** is employed when the objective is to compare the means of two independent groups. For example, one may wish to test whether the mean number of crashes at an intersection before the installation of a speed camera differs significantly from the mean after installation.

The formula for the two-sample t-test is:

$$t = \frac{\overline{X_1} - \overline{X_2}}{s_p\sqrt{\frac{1}{n_1}+\frac{1}{n_2}}} \tag{4.17}$$

Where

X_1, X_2 = sample means of the two groups

n_1, n_2 = sample sizes

S_p = pooled standard deviation, estimated from both groups

The null hypothesis assumes that both groups have the same mean crash frequency ($\mu_1 = \mu_2$). A significant t-statistic ($p < 0.05$) implies that the intervention had a statistically significant effect on crash occurrence.

4.2.7.2 Chi-Square Test

The **chi-square** ($\chi 2$) **test** is designed for categorical accident data, such as crash occurrence frequencies under different roadway or environmental conditions. It is particularly useful for identifying associations between categorical variables—for example, whether the presence of traffic signals is associated with differences in crash frequency at intersections.

The formula for the chi-square statistic is:

$$\chi^2 = \sum \frac{(O-E)^2}{E} \tag{4.18}$$

Where:
O = observed frequency
E = expected frequency under the null hypothesis

If the calculated $\chi 2$ exceeds the critical value from the chi-square distribution (based on degrees of freedom and significance level), then we conclude that there is a significant association between the variables.

4.2.7.3 Analysis of Variance

When comparisons involve more than two groups, the **ANOVA** is more appropriate than multiple pairwise t-tests. ANOVA assesses whether the means of crash data across several groups are significantly different.

The F-statistic is computed as:

$$F = \frac{\text{MS between}}{\text{MS within}} \tag{4.19}$$

Where:

- MS between = variance between group means (explained variance)
- MS within= variance within groups (unexplained variance)

The null hypothesis in ANOVA is that all group means are equal ($\mu_1 = \mu_2 = \mu_3$, etc.). A significant F-statistic ($F > F_{\text{critical}}$ or $p < 0.05$) indicates that at least one group mean differs from the others.

4.2.8 Key Considerations

Assumptions: Many hypothesis tests rely on assumptions such as normal distribution of data, equal variances, and independence of observations. Violation of these assumptions may require the use of non-parametric alternatives (e.g., Mann–Whitney U test, Kruskal–Wallis test). Practical Significance: While hypothesis testing indicates statistical significance, practitioners must also consider the *practical* or *engineering* significance of the results. A small reduction in crashes may be statistically significant but may not justify high intervention costs. Integration with Safety Evaluation: Hypothesis testing is often integrated into before–after studies, matched case-control designs, and experimental evaluations in road safety research, providing robust evidence for policymaking.

4.2.9 Goodness-of-Fit and Model Validation Methods

Once statistical or regression models are developed for accident analysis, it becomes crucial to evaluate their accuracy and reliability. A model that fits the observed crash data well but lacks predictive capability offers little practical value in road safety research or policymaking. Model validation thus serves two main purposes:

1. To assess how well the model represents the underlying crash-generating process.
2. To verify whether the model can be generalized for future or unseen data.

Several widely used goodness-of-fit and validation measures are described below.

4.2.9.1 Coefficient of Determination (R^2)

The coefficient of determination, denoted as R^2, is a fundamental measure of model performance. It quantifies the proportion of variance in the dependent variable (e.g., crash counts or severity) explained by the independent variables (e.g., traffic volume, roadway geometry).

$$R^2 = 1 - \frac{SS_{\text{residual}}}{SS_{\text{total}}} \tag{4.20}$$

- Sum of Squares (SS) SS_{residual}: unexplained variation
- Sum of Squares (SS) SS_{total}: total variation in the data

An R^2 value of 0.8, for instance, implies that 80% of the variability in crashes is captured by the model, while the remaining 20% is influenced by unobserved or random factors. Although useful, R^2 is more appropriate for linear regression models and less reliable for count data models such as Poisson or NB.

4.2.9.2 Akaike Information Criterion

The Akaike Information Criterion (AIC) is a statistical metric that balances model fit with model complexity. It penalizes models that include unnecessary variables, thereby discouraging overfitting.

$$\text{AIC} = -2\ln(L) + 2k \tag{4.21}$$

L: likelihood of the model
k: number of estimated parameters

A lower AIC value indicates a better-fitting and more parsimonious model. For example, when comparing a Poisson model with five variables to an NB model with six variables, the model with the smaller AIC is typically preferred.

4.2.9.3 Likelihood Ratio Test

The likelihood ratio test is employed to compare nested models (i.e., one model is a special case of another). In road safety, this is commonly applied when deciding whether the NB model provides a significantly better fit than the Poisson model.

$$\chi^2 = -2(\text{ln Lrestricted-ln Lfull}) \tag{4.22}$$

If the calculated statistic exceeds the critical chi-square value, then the more complex model (often NB) is statistically justified.

4.2.9.4 Cross-Validation Techniques

Cross-validation provides an empirical way to test predictive accuracy. The dataset is typically divided into two subsets: a training set, used to estimate parameters, and a testing set, used to evaluate predictive performance. K-fold cross-validation further refines this approach by dividing the dataset into k-parts, repeatedly training on k–1 folds and validating on the remaining fold. This ensures robustness and reduces dependence on a single data partition.

4.2.9.5 Prediction Error Metrics: Mean Absolute Error and Root Mean Square Error

Prediction error statistics provide direct measures of how closely model predictions match observed crash frequencies.

Mean Absolute Error: Average absolute difference between observed and predicted values.

Root Mean Square Error (RMSE): Penalizes larger errors more heavily by squaring the differences before averaging.

$$\text{RMSE} = \sqrt{\frac{\sum \left(y_{\text{obs}} - y_{\text{pred}}\right)^2}{n}} \tag{4.23}$$

For instance, if RMSE = 3.2, it means that, on average, the model's crash predictions deviate from actual crash counts by approximately three crashes.

4.2.10 Limitations of Statistical Methods

There are several issues that are as follows: data quality issues of crash reports may have underreporting, missing data, or inconsistent formats. Assumption about the dependent variables and model issues of Poisson assume equal mean and variance; often violated in real data. Omitted Variable Bias consists of some risk factors (e.g., driver fatigue, enforcement intensity) that may not be included. Also, causality problem of statistical associations does not always imply cause-effect relationships. Moreover, sample size sensitivity includes small datasets that may lead to unreliable results. Overfitting risk of models may fit past data well but fail in future predictions.

4.3 ADVANCED METHODS OF ACCIDENT ANALYSIS

In the domain of road safety research, traditional statistical models such as regression and hypothesis testing have historically provided the foundation for accident analysis. While these techniques remain essential, they often rely on restrictive assumptions—such as linearity, independence, or fixed distributions—that may not adequately capture the complex, nonlinear, and heterogeneous naturc of real-world crash data. The growth of computational power, advances in artificial intelligence (AI), and the availability of large-scale traffic and accident databases have encouraged researchers to adopt advanced analytical methods. These methods are not only capable of handling high-dimensional datasets but also reveal subtle patterns in driver behavior, roadway features, and environmental factors that contribute to crash risk. In this section, we examine several advanced techniques, including ML, Bayesian models, spatial and Geographic Information System (GIS)-based approaches, survival analysis, and data mining methods. Together, these represent a shift from descriptive and inferential analysis toward more predictive and decision-support-oriented frameworks in road safety research.

4.3.1 ML and AI in Accident Prediction

ML and AI have gained prominence in transportation safety because of their ability to identify complex, nonlinear relationships in data without relying on predefined parametric forms. Unlike traditional regression models, which assume specific statistical distributions, ML algorithms learn directly from the data and adapt their structure to maximize predictive accuracy.

4.3.1.1 Support Vector Machines

Support Vector Machines (SVMs) are supervised learning methods primarily used for classification tasks. In accident prediction, SVMs separate accident outcomes (e.g., high-severity vs. low-severity crashes) by constructing an optimal hyperplane that maximizes the margin between categories. This makes SVM particularly useful when the relationship between explanatory variables and crash outcomes is not easily represented by linear functions.

- **Mathematical Formulation:**
  - Given dataset (x_i, y_i) where x_i is accident features (e.g., speed, road type) and y_i is the crash outcome:

$$f(x) = w \cdot x + b \tag{4.24}$$

    where w is the weight vector and b is the bias.
  - **Application Example**: Predicting whether a road segment belongs to high-risk or low-risk categories.

4.3.1.2 Random Forest

Random Forests (RFs) are an ensemble method that combines the outputs of multiple decision trees. Each tree is trained on a random subset of data and variables, and the final prediction is made through majority voting (for classification) or averaging (for regression). RFs is well-suited for the analysis of road accident data due to nonlinear interactions of the variables. Also, RFs model provides a measure of variable importance, identifying factors such as driver age, lighting, vehicle type etc. most significant contributing factors on crash severity, and determining the relative contribution of weather, roadway conditions, and traffic density in predicting accident severity.

4.3.1.3 Artificial Neural Networks

Artificial Neural Networks (ANNs) are computational models inspired by the human brain, consisting of interconnected layers of artificial neurons. ANNs can capture highly nonlinear, multidimensional relationships by learning from examples in the training dataset. They have been applied to predict crash frequency, accident severity, and even real-time likelihood of crashes in intelligent transport systems (ITSs). The major advantages of ANN model are flexibility, scalability to large datasets, and adaptability to real-time applications. The major use of ANN model is predicting injury severity categories (minor, serious, fatal) by incorporating multiple risk factors simultaneously.

- **Mathematical Representation:**

$$y = f\left(\sum_{i=1}^{n} w_i x_i + b\right) \tag{4.25}$$

where w_i are weights, x_i are inputs, b is bias, and f is activation function.

4.3.2 Bayesian Models for Road Safety

Bayesian statistics provide a probabilistic framework that is particularly advantageous in accident analysis. Unlike frequentist methods that rely solely on observed data, Bayesian models incorporate both prior knowledge (e.g., historical crash data, expert judgment) and new evidence (e.g., recent crash counts) to update beliefs about risk. The foundation of Bayesian inference is Bayes' Theorem:

$$P(\theta|D) = \frac{P(D|\theta)\cdot P(\theta)}{P(D)} \tag{4.26}$$

where:

$P(\theta|D)$ = posterior probability of model parameters given data
$P(D|\theta)$ = likelihood of data
$P(\theta)$ = prior distribution (belief before data)
$P(\mathrm{D})$ = evidence (normalizing constant)

4.3.3 Spatial Analysis & GIS in Accident Hotspot Detection

Accidents rarely occur randomly; instead, they tend to cluster in specific geographic locations, often referred to as black spots or hotspots. Spatial analysis combined with GISs has become a powerful tool to detect, map, and analyze these clusters. The kernel density estimation produces a smooth density surface, highlighting areas where crashes are concentrated. Spatial autocorrelation (e.g., Moran's I, Getis-Ord Gi*): which tests whether accidents are spatially clustered, dispersed, or randomly distributed. Furthermore, GIS mapping integrates accident records with road geometry, land use, traffic flow, and environmental conditions to provide a contextual view of crash risk. The major applications of spatial analysis are identifying high-priority hazardous locations for intervention. Also, linking accident occurrence to roadway characteristics (e.g., curvature, intersections, signage). It is also helping in monitoring the impact of infrastructure improvements or policy changes on crash distribution.

4.3.4 Survival Analysis for Injury Severity Prediction

Survival analysis is used when the variable of interest is time until an event occurs such as the time from a crash to fatality, or the time between successive accidents on a roadway. Unlike standard regression, survival analysis accounts for censored data (cases where the event has not yet occurred at the time of analysis). The Kaplan-Meier estimator is a non-parametric method that estimates the probability of survival over time. Also, the parametric models (Weibull, exponential, log-logistic) assume specific functional forms for survival times, useful when the hazard rate follows a predictable pattern. Whereas the Cox proportional hazards model is a semi-parametric approach linking survival times to explanatory variables. The hazard function is given as:

$$h(t|X) = h_0(t) \cdot e^{\beta X} \tag{4.27}$$

where $h_0(t)$ is the baseline hazard and X represents covariates (e.g., seatbelt use, emergency response time).

The major applications of these models are estimating the probability of death as a function of response time after serious crashes. Understanding how factors such as safety equipment, medical services, or crash location influence survival chances. Predicting time intervals between successive crashes at hazardous locations.

4.3.5 Data Mining Techniques in Accident Databases

Generally, the data mining techniques focused on the patterns of accidents based on the frequency and correlated factors. The data mining refers to techniques for extracting meaningful patterns, correlations, and associations from large accident datasets that may not be apparent through traditional methods. The following methods are commonly adopted as data mining techniques. The Clustering (K-means, Hierarchical) groups crashes into categories with similar characteristics, such as weather-related vs. alcohol-related crashes. Furthermore, Association Rule Mining (e.g., Apriori algorithm) identifies frequent co-occurrence patterns. Example: *If a crash occurs at night and involves a young driver, then the probability of severe injury is high.* Whereas classification models such as Decision Trees, Naïve Bayes, and SVM assign crashes to categories such as severity levels based on predictor variables. These models reveal hidden relationships among variables (e.g., weather, driver age, time of day). Moreover, these data mining models provide evidence-based support for safety campaigns (e.g., targeting seatbelt use or fatigue awareness). Also, they enhance predictive crash models by uncovering interactions that may not fit linear frameworks.

4.4 CONFLICT ANALYSIS IN ROAD SAFETY

4.4.1 Concept and Importance

Road traffic safety research has long relied on accident statistics to assess the performance of road networks. While crash records provide direct evidence of safety outcomes, they often present serious limitations: underreporting, inconsistency in police records, time delays in availability, and the inability to capture the full scope of risk exposure. To overcome these challenges, researchers and practitioners have turned to conflict analysis a proactive technique that identifies hazardous conditions before actual crashes occur. A traffic conflict is generally defined as a situation where two or more road users are on a collision course and one or more must perform an evasive maneuver (such as braking, swerving, or accelerating) to avoid a crash. These conflicts represent "near-misses," which occur much more frequently than actual accidents. As a result, they provide a rich dataset for evaluating safety at specific sites without the need to wait for serious crashes to happen.

The conflict analysis is a proactive approach; unlike crash-based methods, conflict studies allow identification of risks in advance, making them suitable for evaluating new designs, pilot projects, and short-term interventions. These conflict-based accident approaches facilitate the early detection of hazardous conditions or road accidents, indicating that places may exhibit low collision frequency yet substantial near-miss activity. Also, there are several economic advantages because it reduces dependence on extensive crash records, which are costly and slow to collect. Also, it supports safe system approach by highlighting unsafe interactions between road users. Conflict analysis helps to identify systemic weaknesses in speed management, design geometry, or traffic control.

4.4.2 Traffic Conflict Techniques

The analysis of traffic safety has traditionally relied on crash records; however, accident data often suffer from limitations such as underreporting, lack of detail, and the long-time horizons required to obtain statistically significant trends. To overcome these challenges, researchers and practitioners have developed Traffic Conflict Techniques (TCTs) as proactive methods to assess road safety.

A traffic conflict can be broadly defined as an observable interaction between two or more road users where there is a risk of collision if the participants fail to take evasive action. Unlike accidents, which represent the end stage of this process, conflicts occur far more frequently and can therefore serve as valuable indicators of underlying safety issues. Over the last five decades, several standardized TCTs have been proposed internationally, each with its own conceptual foundation, definition of a conflict, severity measure, and data collection procedure. This section reviews three of the most widely applied techniques: the Swedish TCT (Lund Method), the International Highway Traffic Conflict Technique (IHTCT), and the US Federal Highway Administration (FHWA) approaches.

4.4.2.1 Swedish TCT (Lund Method)

The Swedish TCT, also known as the Lund Method, was developed during the 1980s by Christer Hydén at Lund University. It is one of the most influential conflict observation methodologies in Europe and has been applied in numerous empirical studies. According to Hydén, a conflict occurs when two road users are on a collision course, and the continuation of their current movements would result in an accident unless at least one of them performs an evasive maneuver. This definition emphasizes the *potentiality* of an accident rather than the accident itself, making the technique especially useful for preventive safety analysis.

4.4.2.1.1 Severity Measurement

The severity of a conflict is evaluated using two parameters. (i) Time-to-Accident (TA): The estimated time remaining until a collision would occur if neither party changed speed or direction. A shorter TA implies a more serious conflict. (ii) Intensity of Evasive Action: The level of evasive maneuver required (e.g., hard braking, sharp swerving) is considered alongside TA to provide a graded scale of severity. The Swedish method is particularly valued for its detailed severity grading, which allows conflicts to be categorized into multiple levels. This makes it suitable for in-depth research studies and policy-oriented evaluations. Its adoption in various European projects underlines its reliability and academic credibility. The weakness of this method is observer-intensive and requires field personnel to undergo substantial training. Observer subjectivity can influence the judgment of evasive maneuvers, which may affect the consistency of results across different studies.

4.4.2.2 International Highway Traffic Conflict Technique

The International Highway Traffic Conflict Technique (IHTCT) originated in the 1970s under the joint initiative of the Organisation for Economic Co-operation and Development and the IHTCT committee. The main objective of this initiative was to create a globally applicable conflict observation methodology that could be standardized across countries. According to the IHTCT, a conflict is identified based on observable evasive maneuvers such as emergency braking, sudden swerving, or abrupt stopping. The focus is on the *actual behavioral response* of road users when confronted with a potential collision situation. The severity of conflicts of IHTCT reflects variations in the severity of conflicts, and its classification is less comprehensive than the Swedish technique. The severity is generally deduced from the nature and magnitude of the evasive measures used. The primary advantages of IHTCT are its simplicity and versatility, rendering it appropriate for diverse cultural and infrastructural situations. It has been effectively implemented in nations including Canada, Japan, and the United States, demonstrating its global significance. The major constraint of this method is its restricted severity classification. In contrast to the Swedish technique, it lacks a nuanced assessment of conflict severity. Furthermore, like many observational methodologies, outcomes are affected by the subjective perceptions of field observers.

4.4.2.2.1 Surrogate Safety Measures

Conventional crash-based safety analysis techniques predominantly depend on past accident data. Although this data offers direct proof of safety concerns, it is constrained by severe restrictions, such as underreporting, temporal delays, and the necessity for substantial sample sizes before statistically significant trends can be identified. To rectify these deficiencies, researchers have progressively adopted SSMs. SSMs are quantitative metrics that utilize observable traffic variables—such as vehicle trajectories, speeds, accelerations, and spatial gaps—to assess the probability and severity of near-crash incidents. Rather of awaiting incidents, SSMs enable researchers to proactively discover hazardous interactions and evaluate safety in real time. This methodology has been enhanced by developments in sensor technology, video analytics, and microscopic traffic simulation models, establishing SSMs as fundamental to contemporary traffic conflict analysis. Three prominent SSMs include Time-to-Collision (TTC), Post-Encroachment Time (PET), and Deceleration Rate to Avoid Crash (DRAC).

4.4.2.2.1.1 Time-to-Collision

The TTC is the time remain before two vehicles impact, assuming they maintain their current velocities and trajectories without any evasive maneuvers. A lower TTC rating signifies an increased chance of collision. A TTC of less than 1.5 s is typically categorized as a critical conflict, necessitating prompt evasive measures.

$$\mathrm{TTC} = \frac{D}{V} \tag{4.28}$$

Where:

D = distance between the two vehicles

V = relative closing speed (i.e., difference in velocity between the following and leading vehicles)

Example:

Suppose Vehicle A is traveling at 60 km/h and is 25 m behind Vehicle B, which is traveling at 50 km/h.

Relative speed, $V = (60 - 50) = 10\,\text{km/h} \approx 2.78$ m/s

$D = 25$ m

TTC = 25/2.78 ≈ 9.0 s → Safe margin.

If the gap reduces to only 5 m, then:

TTC = 5/2.78 ≈ 1.8 s → Near-critical.

This demonstrates how TTC can dynamically reflect changes in risk levels as spacing and speeds vary.

4.4.2.2.1.2 Post-Encroachment Time

The PET corresponds to the interval between two cars, or a vehicle and another road user like a pedestrian, traversing the same potential conflict location. A higher PET value often signifies a safer encounter. A PET under 1.0–1.5 s is typically deemed risky due to the minimal margin for error it provides.

$$\mathrm{PET} = t_{\text{second}} - t_{\text{first}} \tag{4.29}$$

Where:

t_{first} = time when the first vehicle clears the conflict point

t_{second} = time when the second vehicle arrives at the same point

Example:

Vehicle A passes through a junction at **10:00:05**.

Vehicle B enters the same conflict point at **10:00:07**.

Thus, PET = 2 s, this is a safe interaction, since both vehicles had sufficient temporal separation.

However, if Vehicle B had entered at 10:00:05.8, then PET = 0.8 s, which indicates a dangerous conflict with high crash potential.

4.4.2.2.1.3 Deceleration Rate to Avoid Crash

The DRAC represents the lowest deceleration required by a trailing vehicle to prevent a collision with a leading vehicle, based on their respective speeds and distances. Elevated DRAC values signify hazardous conditions necessitating abrupt braking. Standard passenger vehicles may safely decelerate at around 3.0–4.0 m/s² under typical road conditions. If DRAC surpasses this threshold, collision avoidance may become physically challenging or uncomfortable for drivers.

$$\text{DRAC} = \frac{V^2}{2D} \tag{4.30}$$

Where:

V = relative speed between the two vehicles

D = current spacing (distance headway)

4.4.3 Field Data Collection for Conflict Analysis

The foundation of any traffic conflict study lies in the quality and reliability of the data collected. Since conflicts are dynamic and often subtle events that occur within seconds, their accurate identification requires systematic observation and precise measurement techniques. Over the years, data collection methods have evolved significantly, ranging from traditional manual observation to advanced computational and simulation-based approaches. The choice of method typically depends on factors such as resource availability, technological infrastructure, cost, and the objectives of the study.

a. **Manual Observation**

One of the earliest and simplest approaches to conflict analysis involves **trained observers** stationed at study sites. These personnel record traffic conflicts in real time, classifying them based on evasive maneuvers, relative positions, and severity levels. Observers may use structured forms, tally sheets, or handheld devices to record data. The primary advantages of the manual method are its low cost and the absence of the need for advanced technology. This manual observation technique offers prompt on-site information. This method is advantageous for pilot investigations or areas with restricted technological resources. Several limitations are present, including susceptibility to observer subjectivity, as assessments of evasive maneuvers and conflict severity may differ among persons. Observer weariness diminishes accuracy, particularly in high-traffic intersections. This approach is inappropriate for extensive or prolonged studies because of limitations in personnel resources. Notwithstanding these deficiencies, manual observation continues to be pertinent in underdeveloped nations or small-scale initiatives where sophisticated technologies are not easily accessible.

b. Video Recording and Computer-Based Analysis

The use of video cameras to capture traffic interactions has become a conventional technique in conflict studies. Video data enables researchers to repeatedly analyze traffic interactions, hence enhancing the precision of conflict identification. Furthermore, sophisticated computer vision software can autonomously derive vehicle trajectories and compute surrogate safety metrics. A notable instrument is the Surrogate Safety Assessment Model (SSAM) created by the US FHWA. SSAM evaluates vehicle trajectory data to autonomously calculate metrics like TTC, PET, and DRAC.

The advantages of the video recording method include the creation of permanent records that may be studied repeatedly. It facilitates the objective and automated calculation of SSMs. This technique allows researchers to examine intricate traffic situations that are not readily observable by the human eye. This method's primary shortcomings necessitate high-resolution sensors and substantial data storage capacity. This entails privacy and legal issues, particularly in urban locales with stringent surveillance laws. Moreover, processing video data using trajectory-extraction techniques can be computationally intensive.

c. Drones and Unmanned Aerial Vehicles

Recent years have seen the increasing use of drones and unmanned aerial vehicles (UAVs) for traffic conflict studies. Positioned above intersections or roadway segments, drones capture a bird's-eye view of vehicle interactions, covering wide areas without interfering with normal traffic flow. There are several advantages: it provides high spatial coverage, especially useful at complex intersections and roundabouts; minimally intrusive to traffic operations, as drivers are usually unaware of overhead monitoring; captures precise vehicle trajectories from an aerial perspective, reducing blind spots common with roadside cameras. Whereas some limitations are as follows: data collection by UAV may be restricted by aviation regulations and local airspace rules. It requires skilled operators and battery management for continuous monitoring. Moreover, data processing, particularly for extracting trajectories from aerial video, is computationally intensive. UAV-based data collection has shown great promise in bridging the gap between field observation and simulation, especially when large-scale, high-precision data is required.

d. Microsimulation Models

Besides field-based techniques, researchers frequently utilize microsimulation models like VISSIM, PARAMICS, or AIMSUN. These models emulate traffic at the individual car level, enabling researchers to reproduce and examine traffic interactions under regulated settings. Simulation software is capable of producing virtual vehicle trajectories. These trajectories can be entered into systems such as SSAM to calculate surrogate safety metrics. Various roadway designs, signal timings, and traffic demand scenarios can be simulated electronically, facilitating "what-if" safety studies without the need for physical interventions. It generally obviates the necessity for comprehensive field data collecting. The microsimulation models facilitate the evaluation of prospective situations (e.g., the effects of new infrastructure, autonomous vehicles). Furthermore, it offers a secure and economical method for analyzing infrequent yet high-risk traffic exchanges. Several constraints exist, including the reliance on the calibration and validation of the simulation model against empirical data. This approach typically cannot completely emulate human driving behavior or environmental conditions. It necessitates specialized software and trained workers.

4.4.4 Case Studies

The use of TCTs and SSMs in several countries has yielded significant insights into their efficacy in forecasting collisions and informing safety interventions. This collection of case studies demonstrates the successful application of conflict analysis across several geographical situations.

4.4.4.1 Case Study 1—Sweden (Lund University)

Sweden is widely recognized as the birthplace of systematic conflict analysis. At Lund University, Christer Hydén pioneered the Swedish TCT in the 1980s, applying it extensively at urban intersections. (Hyden, 1987).

- **Methodology:**
  Observers categorized conflicts according to evasive maneuvers and assessed their severity use TA and maneuver intensity.

- **Findings:**
  A significant link was identified between the incidence of recorded conflicts and subsequent crash data at the identical crossings. Locations with elevated conflict intensity were more prone to have accidents over time.

- **Policy Impact:**
  According to the findings, numerous crossings were reconfigured to enhance bike tracks, pedestrian amenities, and signal adjustments. These changes directly corresponded with Sweden's Vision Zero ideology, which prioritizes proactive safety enhancement.

This instance established TCTs as a reliable instrument for predictive safety analysis and a standard for subsequent international applications.

4.4.4.2 Case Study 2—Canada (Toronto)

In Toronto, Canada, researchers applied the IHTCT to examine school zones, where vulnerable road users, particularly children, faced increased risk.

- **Methodology:**
  Observers documented pedestrian–vehicle conflicts based on evasive actions (e.g., braking or swerving) during school peak hours.

- **Findings:**
  Locations with a high risk of pedestrian-vehicle interactions were identified. Significantly, these locations subsequently documented kid collision events, confirming the method's predictive efficacy.

- **Policy Impact:**
  Interventions comprised the implementation of lowered speed restrictions, installation of school zone signage, and assignment of crossing guards. These measures not only augmented safety but also bolstered community trust in conflict-oriented safety research. This instance underscored the practical applicability of conflict analysis in safeguarding vulnerable road users and demonstrated its policy significance in North America.

4.4.4.3 Case Study 3—India (IIT Delhi, 2020s)

In India, where crash data frequently experience underreporting and lack credibility, researchers have utilized UAV-based conflict analysis to deliver more objective safety evaluations. A recent study at IIT Delhi illustrated the utilization of drone-based video data in conjunction with surrogate metrics (TTC, PET, DRAC). Tiwari, G., and Mohan, D. (2021).

- **Methodology:**
  UAVs recorded aerial imagery of traffic interactions at both signalized and uncontrolled crossings. Vehicle trajectories were retrieved and examined with surrogate measurements.

- **Findings:**
  The research indicated that uncontrolled crossings demonstrated roughly thrice the number of conflicts compared to signalized intersections. Conflicts were more prevalent during peak hours, with numerous instances categorized as serious (low TTC, low PET).

- **Policy Impact:**
  Suggestions were proposed for signage, pedestrian crosswalks, and speed-reduction interventions at essential sites. The research illustrated

the efficacy of UAV-based methodologies in data-deficient contexts like India. This study demonstrates the adaptation of sophisticated conflict analysis methodologies to address the distinct constraints faced by developing nations, where traditional crash-based analysis frequently proves to be incorrect.

4.4.5 Critical Reflections

The review of TCTs, SSMs, and field applications indicates that conflict analysis has emerged as a crucial element of proactive road safety management. Nonetheless, its efficacy relies on meticulous evaluation of both advantages and drawbacks.

4.4.5.1 Advantages of Conflict Analysis

In contrast to crash-based methodologies that depend on historical data, conflict analysis detects possible hazards prior to the occurrence of accidents. This is particularly advantageous for preventive measures. Gathering conflict data necessitates shorter observation durations compared to the compilation of statistically significant crash records, rendering it a time-efficient and cost-effective instrument for safety analyses. Conflict analysis directly underpins the Safe System concept, which emphasizes the reduction of both the probability and severity of collisions by concentrating on human constraints and system resilience. Experimental designs, including roundabouts, cycle tracks, or smart crossings, can undergo safety testing via conflict analysis prior to extensive adoption, therefore decreasing the risk of failure.

4.4.5.2 Limitations of Conflict Analysis

Manual conflict approaches are susceptible to perceptual discrepancies among observers, which may result in inconsistent outcomes. Absence of a global standardization despite the existence of numerous TCTs, a consistent definition of conflict or a standardized severity score is lacking, complicating comparative analyses across research. Advanced techniques utilizing video analytics, UAVs, and computer vision necessitate substantial data storage, significant computational power, and specialized knowledge, potentially constraining their implementation in resource-limited environments.

4.5 DATA FUSION AND EXPLORATION IN ACCIDENT ANALYSIS

The examination of road accidents is intrinsically intricate due to the involvement of numerous interrelated components, including driver's behavior, vehicle dynamics, road geometry, and environmental conditions. Historically, accident research has primarily relied on single-source datasets, predominantly police collision reports. Relying on a single data source frequently leads to incomplete, biased, or inconsistent conclusions. For instance, although police files may document crash conditions, they frequently under-represent small collisions; hospital records may provide details on injury outcomes but omit information regarding the origins of the crashes; and in-vehicle GPS may record trajectories without noting medical repercussions. The disparity has resulted in the development of data fusion methodologies in road safety research. Data fusion denotes the methodical amalgamation of diverse accident-related datasets into a cohesive framework that encapsulates the entire "cause-to-consequence" continuum of a collision. By integrating several sources including police records, hospital reports, insurance claims, telematics, GPS, and closed-circuit television (CCTV) data, researchers can achieve a more comprehensive, accurate, and predictive insight into road accidents.

4.5.1 Concept and Need in Accident Analysis

The primary impetus for data fusion in road safety is the acknowledgment that no singular dataset offers a comprehensive overview of road accident occurrences. Each source represents merely a segment of the comprehensive process: police reports emphasize the site of the collision and legal infractions but frequently neglect to include harm consequences. Furthermore, hospitals document medical outcomes but infrequently associate them with the causes of accidents. Insurance records have financial implications but lack behavioral data. The sensors and telematics record driving dynamics but do not document post-crash outcomes.

Consequently, the requirement for data fusion stems from the imperative to guarantee comprehensiveness: fusion encompasses the complete crash lifecycle, from causation to repercussions. Enhancing accuracy through cross-validation among datasets mitigates uncertainty and reporting bias. Additionally, it facilitates timeliness: Internet of Things (IoT)-based telematics provide near real-time surveillance of roadway hazards. Moreover, it augments predictive value through the integration of extensive datasets into ML models, facilitating the proactive detection of vulnerabilities and high-risk behaviors. Data fusion fundamentally converts accident analysis from a reactive, disjointed procedure into a proactive, comprehensive approach consistent with the safe system approach.

4.5.2 Multi-Source Data Integration

Contemporary accident analysis incorporates various heterogeneous data sources, each possessing unique advantages and disadvantages. Successful fusion necessitates synchronization among these channels.

a. Police Data

- Police data is the most accessible and conventionally utilized source. This data encompasses information regarding crash type, location, road shape, and driver infractions. The primary restriction of police data is the underreporting of small accidents and the subjective nature of officers in documenting causes.

b. Hospital and Health Records

- Hospital data encompasses medical outcomes: injury severity, treatment length, and fatality rates. It is crucial for calculating indices such as the Injury Severity Score (ISS). The primary limitation of hospital data is generally not associated with the site of the event or its causes.

c. Insurance and Legal Data

- Insurance and legal data offer insights into the economic ramifications of accidents via claims and compensation. Additionally, it facilitates cost–benefit analysis of safety interventions. Nevertheless, access to insurance-related data is restricted due to privacy considerations.

d. IoT and Telematics Data

- Data for IoT and telematics are obtained from in-vehicle sensors, cellphones, and event data recorders (black boxes). It offers speed profiles, braking intensity metrics, and lane departure alerts. Additionally, it facilitates real-time risk detection and monitoring of driver behavior.

e. GPS and Navigation Systems

- Recent advancements in vehicle data sources, particularly those utilizing GPS technology, provide high-resolution geographical and temporal data regarding vehicle trajectories. It is beneficial for expediting analysis, congestion assessments, and route-risk mapping.

f. CCTV and Video Analytics

- CCTV and video analytics enable visual verification of collision dynamics. Furthermore, computer vision algorithms are capable of identifying near-misses and hazardous behaviors.

4.5.3 Data Cleaning and Pre-Processing Methods

Prior to data fusion, raw datasets must be subjected to cleansing and standardization to eliminate mistakes and inconsistencies. This step is essential, as incorrect data can undermine fusion accuracy.

a. Error Detection and Correction:

- Every method of collecting road accident data has numerous drawbacks. Identifying abnormalities, such as GPS drifts or typographical errors, is vital. The data indicating "Vehicle speed recorded as 500 km/h" has been highlighted as implausible. Both manual and automatic data gathering methods may provide outliers, necessitating data processing to eliminate these findings.

b. Standardization:

- Standardizing units (km/h versus mph) and injury assessment scales, the Abbreviated Injury Scale and the ISS, may encounter various data sourcing issues, necessitating thorough processing of these discrepancies.

c. Deduplication:

- Merging multiple reports (e.g., hospital + police) referring to the same crash event.

d. Missing Data Imputation:

- Numerous techniques exist for identifying missing data, such as the statistical procedures of mean or median substitution. Furthermore, in recent days, ML approaches such as K-Nearest Neighbors and Expectation-Maximization are predominantly utilized.

e. Geocoding and Temporal Alignment:

- A geocoding and temporal alignment technique transforms text-based crash descriptions ("near temple") into GPS coordinates. Additionally, aligning timestamps among police, CCTV, and emergency response records. The pre-processing principle might be encapsulated as: "Input of inferior quality results in output of inferior quality." In the absence of thorough cleaning, even the most sophisticated fusion algorithms yield erroneous outcomes.

4.5.4 Fusion Algorithms

After being cleaned and standardized, accident datasets can be integrated with computational fusion tools, which encompass both statistical and AI-driven techniques.

a. Kalman Filter:

- The Kalman filter is a widely utilized approach and a recursive algorithm that guesses unknown variables from noisy data. For instance, integrating GPS, which may experience drift in tunnels, with inertial sensors to achieve precise trajectories. The primary advantage is that it facilitates real-time monitoring of crash risk.

b. Dempster–Shafer Theory of Evidence:

- Dempster–Shafer theory is a probabilistic approach for integrating uncertain or partial evidence. For instance, the police report states: "Cause of the crash = speeding (60% confidence)." The GPS indicates: "Speed exceeds limit (70% confidence)." The increased collective confidence that speeding was the cause of the crash.

c. Bayesian Data Fusion:

- Bayesian data analysis employs conditional probability to integrate prior knowledge with empirical evidence. For example, assessing the probability of collisions in areas with limited data.

d. Multi-Layer Neural Fusion:

- The Multi-Layer Neural technique is an AI-based approach that amalgamates several high-dimensional datasets, including photos, text, and sensor logs. It integrates CCTV footage, GPS data, and police text reports for automated crash classification.

4.5.5 Visualization and Exploratory Data Analysis

The visualization of accident data is a crucial component, as it reveals erroneous datasets, outliers, and the distribution of samples. Upon fusion, accident datasets are analyzed through visualization approaches that elucidate concealed patterns, black spots, and risk trends.

a. Heatmaps:

- Heatmaps are mostly utilized for the visual depiction of crash prevalence across geographic areas. These are effective for pinpointing black spots on highways and urban areas.

b. Network Diagrams:

- Network diagrams illustrate accident sites within roadway systems. Furthermore, it illustrates the clustering of incidents and their spillover effects on neighboring networks.

c. Temporal Graphs:

- Temporal graphs depicting crash frequency by time of day, day of the week, or season are crucial for comprehending the kind and circumstances of crashes. Identifying peak-risk hours, such as late-night speeding, necessitated the analysis of morning pedestrian conflicts to comprehend the severity of crashes among road users.

d. Interactive Dashboards:

- The GIS-enabled technologies employed by transportation authorities facilitate an interactive approach for visualizing and comprehending crash data. For instance, India's Integrated Road Accident Database site amalgamates datasets from law enforcement, transportation, and healthcare, displaying them in real-time dashboards for decision-makers. The exploratory data analysis converts unrefined integrated datasets into decision-making instruments, assisting authorities in allocating resources to high-priority locations and times.

4.5.6 Future Trends in Data Fusion for Road Safety

Future advances in accident data fusion are progressing toward real-time, automated, and predictive applications. Current trends encompass:

1. **Big Data and Cloud Platforms:**
    - Extensive integration of terabytes of IoT, GPS, and CCTV data on cloud servers for comprehensive national monitoring.

2. **AI-Driven Fusion:**
    - Deep learning models proficient in the simultaneous analysis of video streams, black box data, and medical reports.

3. **Blockchain-Based Data Sharing:**
    - Secure, privacy-preserving platforms facilitating transparent data sharing across law enforcement, healthcare institutions, and insurance providers.

4. **Digital Twins of Road Networks:**
    - Virtual replicates of metropolitan road networks that integrate actual crash data with microsimulation models for predictive safety evaluation.

5. **Proactive Safety Alerts:**
    - Vehicle-to-Everything (V2X) fusion systems facilitating real-time communication of crash-risk signals among vehicles.

6. **Integration with Smart Cities:**
    - Integrating crash detection and conflict analysis with ITSs and urban surveillance systems.

4.5.7 Summary

Accident analysis serves as a crucial aspect of research in road safety and traffic engineering. It encompasses the methodical gathering, processing, and interpretation of data related to crashes in order to uncover trends, ascertain underlying causes, and assess the success of safety measures. Given that traffic collisions rank among the top causes of injuries and fatalities globally, the examination of accidents is essential not just as a scientific pursuit but also as a critical policy measure to lower societal costs, enhance infrastructure, and safeguard vulnerable road users. The field of accident analysis is fundamentally multidisciplinary, integrating engineering, statistics, behavioral sciences, and data science. The aim is to progress beyond anecdotal accounts of crashes toward decision-making grounded in evidence, where interventions are supported by thorough statistical assessment. For example, determining whether a specific intersection experiences crash rates that are significantly higher than the regional average necessitates the use of proper statistical methods instead of relying on subjective opinions.

BIBLIOGRAPHY

Ahmed, S., Hossain, M. A., Ray, S. K., Islam Bhuiyan, M. M., and Sabuj, S. R. (2023). "A study on road accident prediction and contributing factors using explainable machine learning models: Analysis and performance." *Transportation Research Interdisciplinary Perspectives*, 19, 100814, ISSN 2590-1982, https://doi.org/10.1016/j.trip.2023.100814

Hyden, C. (1987). *The Development of a Method for Traffic Safety Evaluation: The Swedish Traffic Conflicts Technique. No. 70*, Bulletin Lund Institute of Technology.

Lai Zheng, Tarek Sayed, and Fred Mannering. (2021). "Modeling traffic conflicts for use in road safety analysis: A review of analytic methods and future directions." *Analytic Methods in Accident Research*, 29, 100142, ISSN 2213–6657, https://doi.org/10.1016/j.amar.2020.100142

Mathew, T. V. (2025, November 14), Accident Studies, https://www.civil.iitb.ac.in/tvm/nptel/582_Accident/web/web.html#x1-130004

Pignataro, L. J. (2019). *Traffic engineering: theory and practice,* Prentice-Hall.

Scholl, L.; Elagaty, M.; Ledezma-Navarro, B.; Zamora, E.; Miranda-Moreno, L. (2019) *A Surrogate Video-Based Safety Methodology for Diagnosis and Evaluation of Low-Cost Pedestrian-Safety Countermeasures: The Case of Cochabamba, Bolivia.* Sustainability, 11, 4737.

Tiwari, G., and Mohan, D. (2021). *Transport and Safety: Systems, Approaches, and Implementation.* Springer Transactions in Civil and Environmental Engineering, Springer, Singapore. https://doi.org/10.1007/978-981-16-1115-5

5 Surrogate Safety Measures

5.1 INTRODUCTION TO SURROGATE SAFETY MEASURES

Accident frequency and severity are direct measures of road safety, and accident data is generally used to analyze road accidents. But accidents are rare events and sometimes collection of sufficient accident data is not possible. In the absence of accident data, surrogate safety measures (SSMs) are used to predict and analyze the frequency and severity levels of possible potential traffic conflicts using videography data. SSMs are indirect and complementary safety measures and they don't rely on accident data. SSMs are more proactive, more accurate, more informative, and more time-efficient (Hyden, 1987).

Traditional crash-based assessments analyze past accidents to evaluate road safety but have limitations. They rely on historical data, leading to reactive measures, and often face issues with data accuracy and completeness. Low-frequency crash sites and near-miss incidents are often overlooked, limiting a comprehensive safety evaluation. Surrogate safety analysis, characterized by indirect measures such as near-misses and traffic conflicts, stands out as a promising approach to proactively assess and enhance road safety.

There will be a drastic shift of the paradigm by utilizing proactive safety measures or indicators that replace actual safety incidents. By tapping into advanced technologies such as connected vehicles, sophisticated sensor systems, and machine learning algorithms, the aim is to augment the precision and effectiveness of SSMs. Extensive research has been conducted on safety analysis using SSMs. Earlier studies have typically focused on specific themes within safety analysis, such as microsimulation and traditional conflict analysis. While some others have included safety analysis for vulnerable road users, there is a notable gap in comprehensive studies that encompass multiple themes. A comparison of traditional and safe system approach (SSA) is shown in Table 5.1.

5.2 SAFE SYSTEM APPROACH

The SSA emerged in the 1990s from Sweden's Vision Zero and the Netherlands' Sustainable Safety, reframing road safety around the ethical premise that no loss of life is acceptable and that transport systems must be designed for human fallibility and limited biomechanical tolerance to crash forces. Early policies shifted focus from preventing all crashes to preventing deaths and serious injuries by managing kinetic energy through safe speeds, forgiving infrastructure, safer vehicles, supportive user environments, and strong post-crash care, underpinned by shared

DOI: 10.1201/9781003604761-5

TABLE 5.1
Comparison of traditional and safe system approaches

Approach	Traditional Safety Approach	Safe System Approach
Type of crashes	All types of traffic crashes	Fatal and major injury crashes only
Objective	Reducing fatal and injury crashes	Eliminating fatal and injury crashes
Plan of action	Incremental approach to reduce the number of accidents	Proactively targeting and systematic approach
Cause of action	Road users not obeying traffic rules	Road users do make mistakes and users are fragile in nature
Ultimate responsibility	Individual road user	Shared by individuals with system designers
Working of system	Isolated and individual interventions based on necessity	Combination of approaches with a view that even in case one fails, other will provide a safer system

responsibility and proactive governance. By 2021, the World Health Organization (WHO)/UN Global Plan for the Decade of Action for Road Safety centered SSA as the global pathway to reduce road deaths and injuries by 50% by 2030, moving countries from isolated projects to system-wide programs guided by safety performance indicators (SPIs).

5.3 THE FIVE PILLARS OF THE SAFE SYSTEM APPROACH

The SSA was coined based on the five pillars, namely safe vehicles, safe speeds, safer roads, safer people, and post-crash care, and they are detailed as shown in Figure 5.1.

1. Safe Speeds: Set and manage speeds so that, if a crash occurs, the forces remain within human tolerance. This includes designing roads for appropriate operating speeds and using enforcement and technology to sustain compliance.
2. Safe Roads and Infrastructure: Create forgiving, self-explaining environments that separate conflicts, protect vulnerable users, and mitigate crash severity (e.g., median barriers, protected junctions, safe crossings, forgiving roadsides).
3. Safe Vehicles: Increase crash avoidance (e.g., Autonomous Emergency Braking (AEB) and Electronic Stability Control (ESC)) and crashworthiness so both the likelihood and severity of crashes are reduced.
4. Safe Road Users: Support and incentivize protective behaviors (e.g., helmets, restraints), address impairment and distraction, and use education, enforcement, and design to make safe choices easy.
5. Post-Crash Care: Ensure rapid emergency response, trauma systems, and data feedback loops to minimize harm and drive continuous safety improvement.

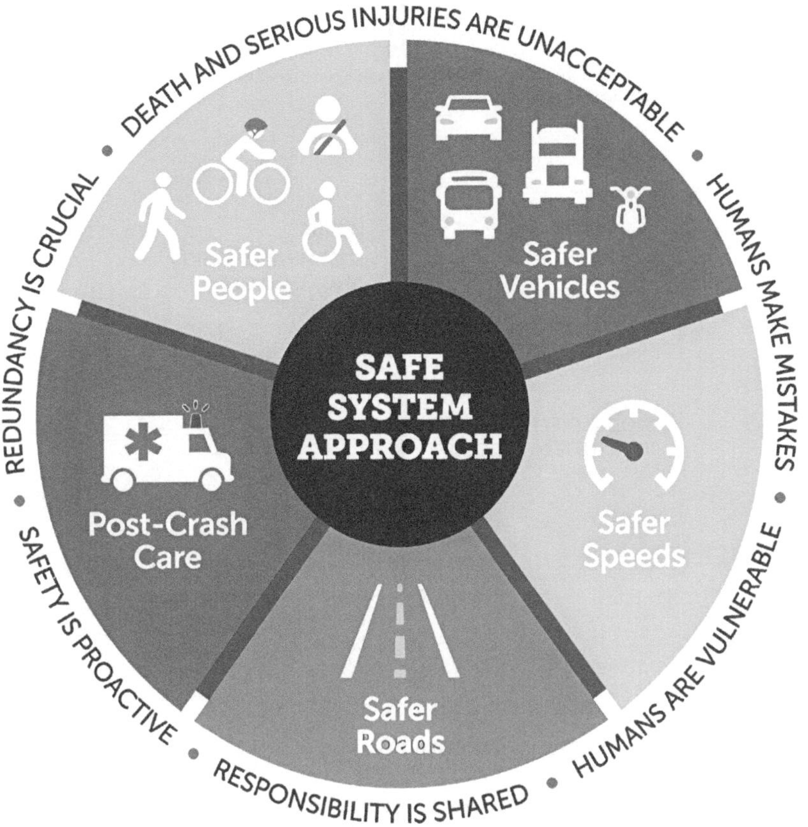

FIGURE 5.1 Five pillars of safe system approach

5.4 EVOLUTION OF SAFE SYSTEM APPROACH

The SSA has evolved over decades as shown in Figure 5.2 and currently, the United Nations second decade of action plan on road safety targets at least a 50% reduction in deaths and injuries by 2030.

5.4.1 1997: Vision Zero Adopted by the Swedish Parliament

Vision Zero was formally adopted by Sweden's Parliament in October 1997, establishing an ethical foundation that no loss of life in road transport is acceptable and that responsibility is shared between system designers and road users. This marked a paradigm shift from preventing all crashes to preventing deaths and serious injuries by ensuring that crash forces remain within human biomechanical tolerance.

Vision Zero reframed safety as a design responsibility: infrastructure, speeds, vehicles, and operations must anticipate human errors and make them non-lethal.

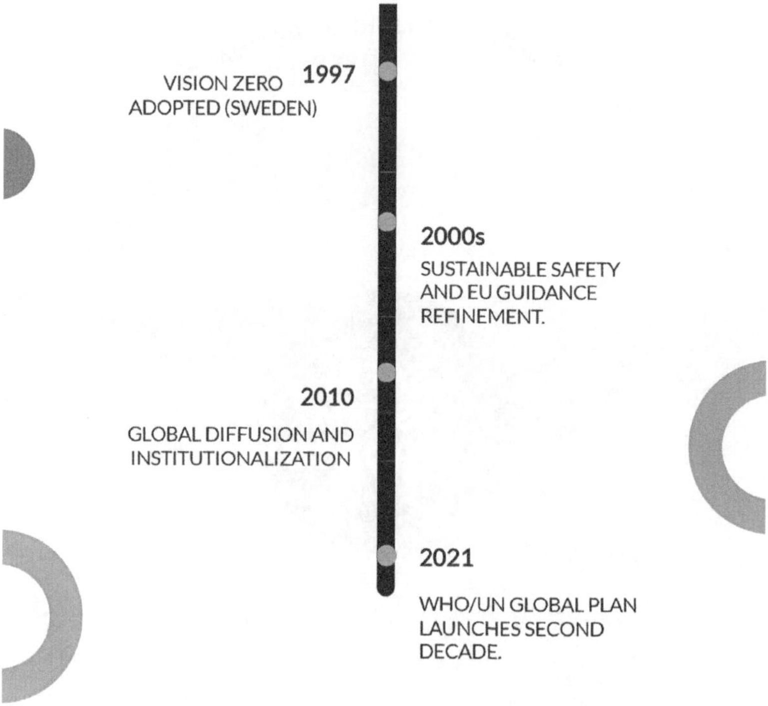

FIGURE 5.2 Timeline of safe system evolution

5.4.2 2000s: Sustainable Safety and EU Guidance Refinement

In parallel, the Netherlands' Sustainable Safety advanced similar system-based thinking: design for human limitations, separate traffic by speed and function, and prioritize forgiving infrastructure. European institutions and expert bodies began codifying proactive risk management, encouraging speed management, intersection safety, and the use of SPIs to manage risk before crashes occur.

This decade built the technical toolbox "force management" via safe speeds, physical separation, and safer roadsides, linking biomechanics to everyday design and operations.

The major focus is related to

- Shift from "black spot" fixes to systemic survivability design.
- Embeds evidence-based treatments (barriers, roundabouts, protected crossings) as standard practice.

5.4.3 2010s: Global Diffusion and Institutionalization

Safe System principles spread beyond early adopters, informing national strategies, regional programs, and city policies across Europe, Australasia, and beyond.

International guidance matured around governance, shared responsibility, and comprehensive delivery across safe roads, speeds, vehicles, users, and post-crash care.

The emphasis shifted to building institutions that can plan, fund, monitor, and continuously improve SSA implementation using SPIs and results frameworks.

5.4.4 2021: WHO/UN Global Plan Launches the Second Decade of Action (2021–2030)

In October 2021, the WHO/UN launched the Second Decade of Action for Road Safety 2021–2030, targeting at least a 50% reduction in deaths and injuries by 2030.

The WHO Global Plan places the SSA at the core of this target, setting a global expectation to implement SSA across five domains: safer roads, speeds, vehicles, users, and post-crash care, supported by governance and monitoring.

Speed management is highlighted as a critical cross-cutting element because of its direct role in controlling kinetic energy and injury severity.

5.5 APPLICATION OF SSA BASED ON FIVE PILLARS

5.5.1 Safe Speeds

Set survivable speed limits by conflict type (e.g., 30 km/h where vehicles mix with pedestrians; lower intersection approach speeds), and redesign streets so operating speeds match targets, not just the posted limit.

Run a speed management program: establish a vision, collect speed/safety data, proactively prioritize sites, implement measures (raised crossings, lane narrowing, cameras), and monitor/evaluate continuously.

Implement area-wide "slow zones" with self-enforcing design (e.g., 20 mph/30 km/h zones) that have shown large reductions in fatal and serious injuries in international evaluations.

5.5.2 Safe Roads and Infrastructure

Add median barriers or convert to 2 + 1 roads on undivided high-speed corridors to prevent head-on crashes and keep impact forces within human tolerance.

Replace high-speed, high-angle intersections with roundabouts or protected turn phases to reduce conflict angles and speeds, lowering severe injury risk.

Provide forgiving roadsides, protected crossings, and separated cycleways to reduce exposure and mitigate severity for vulnerable users.

5.5.3 Safe Vehicles

Accelerate fleet safety via ESC, AEB, lane-keeping, and crashworthiness standards; use New Car Assessment Program (NCAP) ratings and public procurement to shift markets toward safer vehicles.

Integrate vehicle tech with speed and road design (e.g., Intelligent Speed Assistance (ISA) paired with survivable speed limits) to manage kinetic energy at scale.

5.5.4 Safe Road Users

Increase helmet/seatbelt/child restraint use through design that supports safe choices, targeted enforcement, and education campaigns aligned with system changes.

Address impairment and distraction with deterrence, technology, and environmental design that reduces high-risk choices and exposure.

5.5.5 Post-Crash Care

Improve emergency medical service (EMS) response times, dispatch coverage (universal call numbers), trauma capacity, and hospital coordination to reduce mortality and serious harm after crashes.

Establish feedback loops from EMS/hospitals/police data to guide proactive risk management and investment decisions.

5.6 CLASSIFICATION OF SURROGATE SAFETY MEASURES

Typically, surrogate safety methods can be classified as time-based, exposure-based, and severity-based measures.

5.6.1 Time Based

These measures are based on measuring the interaction between vehicles or vehicles and pedestrians using a proximity measure to a collision. For example, in the case of a vehicle crossing an interaction, the time gap between passing of subject vehicle and other vehicles from the adjacent leg of the intersection can be considered as a measure of surrogate safety. In this case, there is a close chance of collision that has been avoided and a minute difference in time might have resulted in a possible accident. There are several measures widely used as time-based parameters such as Time-to-Collision (TTC), Post-Encroachment Time (PET), and Distance to Collision (DTC).

5.6.2 Exposure Based

The parameters based on likelihood of a collision can be classified under exposure-based SSM. These measures rely on the chances of occurrence rather than severity of crash in case of a crash possibility.

5.6.3 Severity Based

Vehicles' paths and speeds are important for proper interpretation of the collision closeness and the severity of potential collision. In other words, the speeds of vehicles during a surrogate event must be sufficiently high to constitute a dangerous outcome of the potential collision and to indicate that the event resulted from human error. Some studies combine proximity with the speeds of the vehicles to arrive at measures that combine the closeness to the potential collision and the severity of the outcome of the potential crash.

Another type of classification generally considered is as follows:

Temporal Proximity Measures: These measures, including TTC and PET, assess how closely vehicles approach each other in time.

Spatial Proximity Measures: Metrics such as Proportion of Stopping Distance and Distance to Intersection analyze the spatial proximity of vehicles, offering a comprehensive view of conflicts.

Kinematic Measures: These involve parameters like acceleration and deceleration, with a focus on heavy braking as an indicator of conflicts.

Mix of Measures: Some methods combine multiple conflict indicators, such as aggregated cross-tendency measures, which blend TTC, reaction time, and deceleration rate.

Combination of Measures: Researchers have used various combinations of conflict measures to capture diverse safety aspects during traffic interactions. These classifications help in comprehensive conflict analysis.

Quantitative Analysis: Surrogate safety analysis has gained widespread traction globally, reflecting its pivotal role in bolstering road safety efforts and its adaptability to diverse traffic scenarios and infrastructure contexts.

Thematic Analysis: Perkins and Harris, while working as engineers at General Motors in 1967, developed the Traffic Conflict Technique as a means to identify safety hazards related to vehicle construction. Their pioneering research led to a significant revelation—they found a strong correlation between conflict patterns on the road and the types of accidents that occurred. Surprisingly, they concluded that the prevalence of conflicts among road users could serve as a better predictor of risk than simply looking at accident rates alone. To define conflicts in a traffic context, Hydén (1987) proposed a straightforward yet insightful perspective: conflicts are situations in which two or more road users would collide if none of them took evasive actions. This definition hinges on the concept of TTC, where the values of TTC and vehicle speeds are used to gauge the severity of these conflicts. The basic definitions of widely used SSMs to predict the frequency and severity levels of possible potential traffic conflicts were discussed below. A brief overview of these measures and their possible applications is highlighted in Table 5.2.

TABLE 5.2
Snapshot of surrogate safety measure details

Measure	Details	Units
Time-to-Collision (TTC)	Expected time for two vehicles to collide if they remain at their present speed and on the same path	Seconds
Post-Encroachment Time (PET)	Time lapse between end of encroachment of turning vehicle and the time that the through vehicle actually arrives at the potential point of collision	Seconds
Gap time	Time lapse between completion of encroachment by turning vehicle and the arrival time of crossing vehicle if they continue with same speed and path	Seconds
Deceleration Rate to Avoid Collision (DRAC)	Rate at which crossing vehicle must decelerate to avoid collision	m/s^2
Proportion of Stopping Distance (PSD)	Ratio of distance available to maneuver to the distance remaining to the projected location of collision	No units
Initially Attempted Post-Encroachment Time (IAPT)	Time lapse between commencement of encroachment by turning vehicle plus the expected time for the through vehicle to reach the point of collision and the completion time of encroachment by turning vehicle	Seconds
Max Speed Difference (ΔV)	The relative speed between two conflicting vehicles	m/s

i. **Time to collision**: It was introduced by Hayward in 1971. It is defined as the time to collision (seconds) when two vehicles continue their trajectory at the same speeds and the same angle without any kind of evasive behavior (Hayward, 1971). Figure 5.3 shows the pictorial representation of TTC.

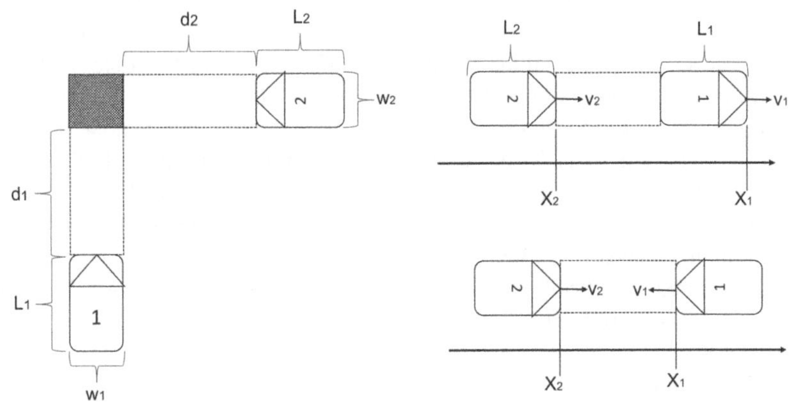

FIGURE 5.3 Pictorial representation of Time-to-Collision (TTC) definition

$$\mathrm{TTC} = \frac{d_2}{v_2} \text{ if } \frac{d_1}{v_1} < \frac{d_2}{v_2} < \frac{d_1 + l_1 + w_2}{v_1} \quad (5.1)$$

$$\mathrm{TTC} = \frac{d_1}{v_1} \text{ if } \frac{d_2}{v_2} < \frac{d_1}{v_1} < \frac{d_2 + l_2 + w_1}{v_2} \text{ (side)} \quad (5.2)$$

$$\mathrm{TTC} = \frac{X_1 - X_2 + l_1}{v_1 - v_2} \text{ if } v_2 > v_1 \text{(rear end)} \quad (5.3)$$

$$\mathrm{TTC} = \frac{X_1 - X_2}{v_1 + v_2} \text{ (head on)} \quad (5.4)$$

Where, d_1, v_1, X_1, and w_1 are the distance from the conflict point, speed, and width respectively for road user 1 and d_2, v_2, X_2, and w_2 are the distance from the conflict point, speed, and width respectively for road user 2.

ii. **Post- encroachment time**: It is the time difference between the first road user leaving the conflict area and the second road user entering the conflict area (Allen, 1978). Figure 5.4 shows the pictorial representation of PET.

Let t_1 be the time at which the first road user is leaving the conflict area and t_2 be the time for the second road user to reach the conflict area.

$$\mathrm{PET} = t_2 - t_1 \quad (5.5)$$

iii. **Deceleration to safety time**: It is the necessary deceleration to reach a non-negative PET value if the movement of the conflicting road users remains unchanged (Hupfer, 1997).

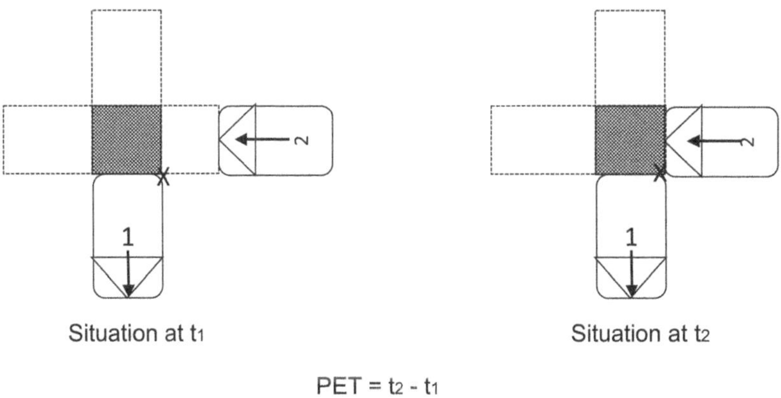

FIGURE 5.4 Pictorial representation of Post-Encroachment Time (PET) definition

iv. **Gap time**: It is the time lapse between the completion time of encroachment by one road user and the arrival time of the interacting road user if they continue with the same speed and path (Archer, 2004).
v. **Risk indicator**: It is derived using PET and approaching speeds of vehicles by Scholl et al. (2019). It is defined as the ratio of approaching vehicle speed over PET.

$$\text{Risk Indicator} = \frac{\text{Approaching Vehicle Speed}}{\text{Post Encroachment Time}}$$

5.7 TYPICAL FIELD EXAMPLES

As outlined in the previous section on SSMs, traffic safety analysis uses quantitative metrics to assess potential conflicts and predict hazards before actual collisions occur. This will enable proactive safety interventions to quantify "near miss" scenarios that objectively support evidence-based intersection design considerations.

A typical example of measuring the TTC and PET values at an intersection is outlined below: determination of PET from a video data collected at a typical roundabout in India is as follows. PET is the time elapsed between the moment the first road user leaves a conflict point and the moment the second road user arrives at the same conflict point.

Mathematically PET can be expressed as $PET = t_2 - t_1$, where t_1 = time when first vehicle leaves the conflict point and t_2 = time when second vehicle arrives at the same conflict point, and this is pictorially represented in Figure 5.5a and 5.5b.

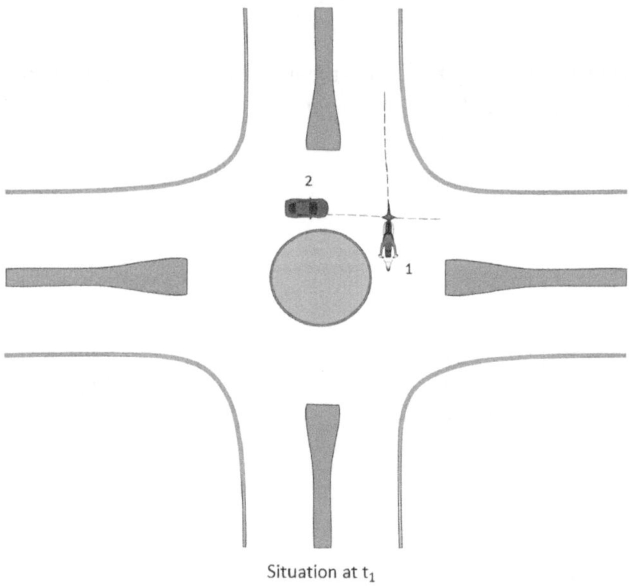

FIGURE 5.5A Pictorial representation of Post Encroachment Time (PET). Time when first vehicle leaves conflict point

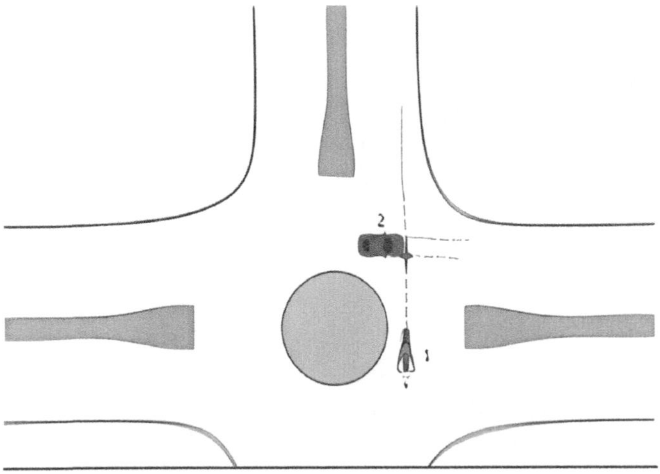

FIGURE 5.5B Pictorial representation of Post Encroachment Time (PET). Time when second vehicle arrives conflict point

The PET values will be collected for all such kind of vehicular interactions and the risk index will be typically calculated as follows. The threshold values for this kind of classification depend on the type of road and intersection, geometry, type of vehicles, traffic conditions, and road user behavior.

- PET < 1 s can be considered as a high collision risk
- 1 ≤ PET ≤ 3 can be considered as a moderate risk
- PET > 3 s can be considered as a low risk

Step-by-Step Procedure

- Identify the conflict point: The spatial location where two vehicle paths intersect
- Measure t_1: Exit time: Record when the first vehicle's last part exits the conflict point
- Measure t_2: Arrival time: Record when the second vehicle's first part reaches the conflict point
- Calculate the difference: Subtract t_1 from t_2 to obtain PET value

The detailed TTC calculation is represented through Figure 5.6 for a realistic understanding of the process and its application.

TTC is an SSM that estimates the time remaining before two vehicles would collide if they continued on their current trajectories and maintained their current speeds.

$$TTC = d / v$$

FIGURE 5.6 Demonstration of Time-to-Collision (TTC) with the help of field example

Where d = distance to the conflict point, v = instantaneous speed of the vehicle.

Minimum TTC (TTC_{min}). The smallest TTC value observed during the entire interaction between two road users.

- $TTC_{min} < 1.0$ s: High collision risk
- $1.0\ s \leq TTC_{min} \leq 1.5$ s: Moderate collision risk
- $TTC_{min} > 1.5$ s: Low collision risk

5.7.1 Summary

The traditional methods of road safety enhancement were heavily dependent on post-response system and remedial measures were taken up after the occurrence of road accidents. In the recent past, SSA is a kind of proactive approach where the chances of occurrence of accidents can be predicted through SSMs. The safe system has been widely implemented in Europe and Australian countries and has the merit of saving precious lives by preventing the occurrence of accidents. The broad introduction to SSA and key indicators of SSMs are presented in this chapter. A field example with a demonstrated calculation of typical SSMs is presented toward the end of this chapter.

BIBLIOGRAPHY

Allen, B. L., Shin, B. T., & Cooper, P. J (1978). *Analysis of traffic conflicts and collisions,* Transportation Research Record, 667, 67–74.

Archer, J., (2004). *Methods for the Assessment and Prediction of Traffic Safety at Urban Intersections and Their Application in Micro-simulation Modelling*, Royal Institute of Technology.

European Commission, European Road Safety Observatory. (2023). *Road Safety Thematic Report: Safe System Approach*. Publications Office of the European Union, Brussels.

International Road Assessment Programme (iRAP). (2022). *Safe System Approach*. iRAP.

International Transport Forum/OECD. (2022). *The Safe System Approach in Action: An Experience-Based Guide to Enhanced Road Safety*. OECD Publishing.

Hayward, J. C. (1971). *Near misses as a measure of safety at urban intersections*, M.S. Thesis, The Pennsylvania State University, Philadelphia, USA.

Hupfer, C., (1997). *Computer Aided Image Processing to Modify Traffic Conflicts Technique*, Transportation Department, University of Kaiserslautern, Kaiserslautern.

Hydén, C. (1987). *The Development of a Method for Traffic Safety Evaluation: The Swedish Traffic Conflicts Technique. No. 70*, Bulletin Lund Institute of Technology.Scholl, L.; Elagaty, M.; Ledezma-Navarro, B.; Zamora, E.; Miranda-Moreno, L. (2019) *A Surrogate Video-Based Safety Methodology for Diagnosis and Evaluation of Low-Cost Pedestrian-Safety Countermeasures: The Case of Cochabamba, Bolivia*. Sustainability, 11, 4737.

PIARC (World Road Association). (2024). *Road Safety Manual: Safe System Elements*. PIARC.

World Bank, Global Road Safety Facility. (2023). *Road Safety Management Capacity Reviews and Safe System Projects: Guidelines for Conducting Reviews and Implementing Projects*. World Bank.

World Health Organization. (2021). *Global Plan for the Decade of Action for Road Safety 2021–2030*. World Health Organization.

6 Surrogate Methods of Road Safety

6.1 INTRODUCTION

In Indian condition most of the times accidents remain unreported especially if that is minor injury or property damage. Major injury also may be unreported sometimes as general public tendency. Different socio-political factors may act behind this. So, the field traffic accident data underestimate real traffic safety concerns. Along with the ambiguity in crash data the availability of data for research work is also tedious and time consuming. Therefore, instead of field crash data, surrogate safety measures (SSMs) are to be used to compute number of critical conflicts. In accordance with several literatures Post-Encroachment Time (PET) and Time-to-Collision (TTC) are found to be the most suitable surrogate safety tool.

The time taken by the following vehicle to collide with the leading vehicle if the speed and direction of the vehicle remained constant. In traffic safety, it is used to estimate the time it will take for an object (vehicle or pedestrian) to collide with another object or a surface. The calculation of TTC depends on the relative motion and distance between the objects. Lower TTC value indicates lesser time requirement to collide with the leading vehicle, i.e., higher risk of collision.

The time difference between the exits of first vehicle to entry of front end of second vehicle at an intersection point. PET indicates the chance of occurrence of accident. If the PET value is high there is more time gap between the vehicles passing same point, which means lesser chance of accident, i.e., less severe conflict.

The major tools and techniques widely used for surrogate safety measurement and analysis are presented in the next section.

6.2 TOOLS AND TECHNIQUES USED FOR SURROGATE SAFETY ANALYSIS

6.2.1 Video-Based Surrogates

Video-based surrogate method offers a valuable tool for transportation professionals, relying on detailed video footage to provide insights into traffic movements and potential safety issues. The merits of this approach lie in its ability to capture rich, visual data that enables a more comprehensive understanding of traffic dynamics. Video-based surrogates aid in analyzing near-misses and critical events for road safety but are hindered by limited camera views, potentially missing crucial incidents. Nevertheless, they provide valuable data for improving traffic safety strategies.

DOI: 10.1201/9781003604761-6

6.2.2 Connected Vehicle Data

Connected vehicle data emerges as a powerful tool in modern transportation systems, leveraging real-time insights to provide a dynamic perspective on traffic patterns and enabling the early detection of safety issues. The merits of this approach are evident in its ability to offer timely and relevant information, facilitating proactive responses to potential hazards. By relying on data generated from connected vehicles, transportation authorities can comprehensively understand road conditions and identify safety concerns in real-time. However, the effectiveness of connected vehicle data is contingent on widespread vehicle connectivity, which may pose challenges in certain regions or contexts where such technology adoption is limited. Despite this dependency, the early detection and real-time analysis of safety concerns through connected vehicle data remain crucial in enhancing overall traffic management and ensuring the safety of road users.

6.2.3 Microsimulation Models

Microsimulation models stand as valuable tools in transportation planning, offering a detailed understanding of traffic dynamics by simulating the movements of individual vehicles. The merits of these models lie in their ability to capture intricate details of traffic behavior, allowing planners to gain insights into the nuances of road systems. By utilizing traffic flow data and vehicle characteristics, microsimulation models enable predictive analysis of safety events in simulated environments, aiding in assessing potential risks and developing targeted mitigation strategies. However, the resource-intensive nature of microsimulation models can limit scalability, making them less practical for large-scale applications. Additionally, the accuracy of model predictions hinges on the quality of input parameters, emphasizing the importance of obtaining precise and reliable data to ensure the fidelity of simulation results. Despite these challenges, microsimulation models play a crucial role in advancing transportation planning and safety analysis by offering a granular perspective on traffic behavior and its implications.

6.2.4 Driving Simulators

Driving simulators serve as invaluable tools in studying driver behavior by providing controlled virtual environments for the examination of specific scenarios. The merits of these simulators lie in their ability to recreate diverse driving conditions, allowing researchers and practitioners to gain insights into how drivers respond to various situations. Researchers can better understand the intricacies of human behavior behind the wheel by utilizing driver behavior data and analyzing simulator outputs. However, a notable demerit of driving simulators is the potential lack of real-world complexity, which may affect the generalizability of findings to actual driving situations. While simulators offer controlled environments for studying specific aspects of driver behavior, the challenge remains in ensuring that the insights gained can be applied to the unpredictability and complexity of real-world driving conditions. Despite this limitation, driving simulators continue to be valuable tools for research, training, and the exploration of driver reactions in controlled settings.

6.2.5 Field Operational Tests

Field operational tests (FOTs) are valuable endeavors in transportation research as they involve the real-world deployment of instrumented vehicles to collect naturalistic driving data. The merits of FOTs are evident in their ability to provide authentic insights into driver behavior and interactions within genuine traffic environments. By observing and analyzing data from instrumented vehicles operating in real-world conditions, researchers gain in-depth knowledge about various aspects of driving behavior, vehicle performance, and interactions with the surrounding environment. However, a notable demerit of FOTs lies in the limited control over variables, as the dynamic nature of on-road scenarios may introduce confounding factors that could affect the accuracy and generalizability of the findings. Despite this challenge, the richness of real-world data collected through FOTs remains invaluable in advancing our understanding of how drivers behave in their natural driving environment, contributing to the development of safer and more efficient transportation systems.

6.2.6 Data Mining Techniques

Data mining techniques play a pivotal role in the analysis of large-scale transportation data by facilitating the identification of patterns and correlations. The merits of these techniques lie in their ability to extract valuable insights from vast datasets, allowing for correlation analysis that can reveal hidden relationships among various variables. In the context of transportation, data mining techniques enable the identification of patterns related to traffic flow, driver behavior, and safety outcomes. However, a notable demerit is the reliance on the quality and availability of data, as the robustness of findings is contingent upon the accuracy and completeness of the input information. Ensuring data quality and addressing potential biases are essential for producing reliable results. Overall, data mining techniques enhance safety predictions and inform evidence-based decision-making in transportation systems by uncovering meaningful patterns within complex datasets.

6.2.7 Machine Learning Algorithms

Machine learning (ML) algorithms stand out as powerful tools in transportation safety analysis, leveraging predictive modeling based on historical data to enhance accuracy in safety predictions. The merits of these algorithms lie in their capacity to uncover complex patterns and relationships within large datasets, enabling more precise forecasting of safety outcomes on the road. By utilizing historical crash data and relevant traffic variables, ML algorithms can identify factors contributing to accidents and assess their impact. However, a notable demerit is the challenge in interpretability, as the intricate nature of some algorithms may make it difficult to understand the underlying decision-making processes. Additionally, the risk of over-fitting, where models become too tailored to historical data and struggle to generalize to new scenarios requires careful consideration. Despite these challenges, ML algorithms play a crucial role in advancing safety predictions in transportation by providing a data-driven approach that can uncover insights for effective preventive measures and policy planning.

6.2.8 Empirical Bayes Methods

Empirical Bayes methods represent statistical approach to estimating crash risk by leveraging observed and expected frequencies. The merits of these methods lie in their ability to provide a robust estimation of crash risk on specific roadways. This provides a nuanced assessment of roadway risks by comparing actual crash rates with expected rates from larger datasets or reference groups. However, a notable demerit is the sensitivity of these methods to small sample sizes, as limited data can impact the reliability of risk estimates. Ensuring an adequate dataset for analysis is essential to mitigate this challenge. Overall, empirical Bayes methods contribute valuable insights into assessing crash risk on specific roadways, aiding in targeted interventions and improvements to enhance overall transportation safety.

6.2.9 Eye-Tracking Studies

Eye-tracking studies serve as a valuable methodology in understanding driver attention by analyzing eye movements in various driving scenarios. The merits of these studies are evident in their ability to provide direct insights into where and how drivers allocate their visual attention. By tracking eye movements, researchers can better understand visual behavior, identifying patterns related to hazard perception, distraction, and information processing. However, a notable demerit of eye-tracking studies is the potential limitation in external validity, as the controlled and artificial conditions of a laboratory setting may not fully replicate the complexities of real-world driving. The impact of these artificial conditions on participant behavior and the generalizability of findings to everyday driving scenarios must be considered. Despite this limitation, eye-tracking studies contribute significantly to our understanding of driver behavior and offer valuable information for designing interventions to minimize distractions and improve overall road safety.

6.2.10 Cognitive Task Analysis

Cognitive task analysis plays a crucial role in exploring the cognitive processes involved in driving tasks. The approach lies in its ability to delve into the mental operations that individuals undergo while performing specific driving-related activities. By examining task-specific data and collecting participant feedback, researchers can identify and understand the cognitive challenges inherent in driving. However, a notable demerit is its susceptibility to participant bias, as self-reported data may not always align with actual cognitive processes. Additionally, the findings may not fully capture the complexities of real-world driving scenarios, as the controlled environment of a study may differ from the dynamic and unpredictable nature of on-road experiences. Moreover, cognitive task analysis remains a valuable tool in uncovering cognitive aspects of driving behavior, contributing insights that can inform training programs, safety interventions, and the design of user-friendly automotive technologies. The discussion highlights the variety of tools and techniques in surrogate safety analysis, chosen based on analysis goals, resources, and safety concerns. Using various methods together gives a thorough understanding for making safe transportation decisions based on evidence. A summary of the various techniques used for surrogate safety analysis with their merits and demerits are highlighted in Table 6.1.

TABLE 6.1
Tools and techniques used for surrogate safety analysis at different traffic facilities

Tool/ Techniques	Merits	Demerits/ Limitations	Data Used	Observations/ Results
Video-based surrogates	Detailed analysis of traffic movements, near-misses	Limited field of view, video footage reliance	Video footage, near-miss observations	Direct traffic behavior insights
Connected vehicle data	Real-time traffic insights, early safety detection	Relies on widespread vehicle connectivity	Vehicle connectivity data	Proactive safety issue detection
Microsimulation models	Simulates individual vehicle movements in scenarios	Resource-intensive	Traffic flow, vehicle characteristics	Predicts safety events with high precision
Driving simulators	Controlled virtual environment for driver behavior	Lacks real-world complexity	Driver behavior data, simulator outputs	Understands driver reactions in controlled settings
Field operational tests	Real-world, naturalistic data	Limited control over variables	Instrumented vehicles, real-world data	In-depth insights into natural driving behavior
Data mining techniques	Identifies patterns in large datasets	Dependent on data quality and availability	Large transportation datasets	Predicts safety outcomes through correlation
ML & DL algorithms	Improves predictive accuracy from historical data	Susceptible to over-fitting, interpretability challenges	Historical crash data, traffic variables	Enhanced prediction of safety outcomes
Empirical Bayes methods	Statistical estimation of crash risk	Sensitive to small sample sizes	Crash data, roadway characteristics	Estimates crash risk on specific roadways
Eye-tracking studies	Analyzes driver attention and distractions	Limited external validity	Eye movement data, driving scenarios	Insights into visual behavior and distractions
Cognitive task analysis	Focuses on cognitive processes in driving tasks	Subject to participant bias	Task-specific data, participant feedback	Identifies cognitive challenges faced by drivers
Hazard rating systems	Systematic assessment of roadway hazards	Subjectivity in hazard ratings	Road infrastructure data, safety audits	Identifies high-risk road elements
Geographic Information Systems (GIS)	Spatial analysis and mapping of traffic data	Limited by data quality and availability	GIS data, spatial features	Identifies spatial patterns and clusters in traffic
Conflict technique	Quantifies potential collision risk	Sensitive to measurement errors	Vehicle trajectories, speed data	Assesses collision risk

Abbreviations: DL: Deep Learning; ML: Machine learning.

6.3 TYPICAL EXAMPLES WITH CASE STUDIES

SSMs or indicators used in various studies. Most of the previous studies used either TTC or PET to analyze the pedestrian-vehicle (P-V) interactions but few studies used other SSMs. The use of either PET or TTC alone is insufficient to measure the injury severity of a potential collision and the inclusion of either pedestrian or vehicle parameters will better predict the potential possible collisions. Most of the other SSMs used for the analysis of P-V interactions were derived from either TTC or PET.

6.3.1 Case Study: Pedestrian Dilemma Zone

One of the applications of SSMs is determining the pedestrian dilemma zone (PDZ) requirements. When a pedestrian is crossing the road, he/she may try to find a sufficient gap and the decision to stop/go depends on the available gap size. But sometimes pedestrians may get confused about whether the available gap is sufficient to cross the road or not. This situation is known as the dilemma of a pedestrian. This situation arises only when the vehicles are present within a certain section of a roadway and this section is known as the PDZ. Gap acceptance behavior is the basic concept for dilemma zone analysis. PDZ is defined as "the section of a roadway where the presence of vehicles creates a stage of confusion for pedestrians while crossing the road" (Pawar et al., 2016). It consists of upper and lower boundary limits and the difference between these two limits is defined as the length of the dilemma zone. Figure 8.1 shows the physical location of a PDZ for one approach near an intersection. D_0 and D_1 are the lower and upper boundary limits respectively within which pedestrians are in a stage of confusion to judge whether the available gap is sufficient or not. There are several methods to estimate the dilemma zone using the filed data and the basic procedures of various methods to estimate PDZ boundary limits were briefly discussed below.

I. **Gap cumulative distribution method**: This method is used to estimate the dilemma zone boundary limits. In this method, a plot between the cumulative percentage gap accepted/rejected on y-axis and distance on x-axis is plotted and then the distance on x-axis corresponding to 10 percentile accepted and the 90 percentile rejected gaps on y-axis are taken as the lower and upper boundary limits of dilemma zone respectively.

II. **Support vector machine (SVM) method**: Used the SVM technique to find the dilemma zone boundaries at an uncontrolled midblock crossing. In this method, a plot between the speeds of vehicles or pedestrians on y-axis and distances of the vehicles from the pedestrian/vehicle trajectory paths on x-axis is plotted on a coordinated system. A hyperplane is constructed as a decision plan to separate the accepted and rejected gap sizes with maximum margin. The lower and upper limits of the dilemma zone are taken on the hyperplane corresponding to the plus or minus two standard deviations from the mean speed.

III. **Binary logit method**: This method has two alternative outputs (accept/reject) as it is a discrete choice model. In this method, a binary logit model is developed using gap accept/reject as a dependent variable and various independent variables, and the lower and upper boundary limits of dilemma zone are estimated corresponding to the 0.1 and 0.9 probabilities.

IV. **Probabilistic method**: This method estimates the dilemma zone boundaries. A detailed procedure to determine the boundary limits of dilemma zone is explained below.
   a. **Classification of speed bands**: First the approaching speeds of vehicles or crossing speeds of pedestrians are classified into bands.
   b. **Setting 10th percentile and 90th percentile boundaries classified speed bands**: Cumulative 10th percentile minimum accepted gaps and 90th percentile maximum rejected gaps are determined for various speed bands. After that, mark the midpoint of each speed band corresponding to the 10th percentile minimum accepted gaps and 90th percentile maximum rejected gaps.
   c. **Plotting profile**: 10th and 90th percentile profiles for the midpoints are fitted and then lower and upper boundary limits of dilemma zone are taken on x-axis by projecting the average speed over the profiles.

Despite some fluctuations, pedestrian deaths are increasing every year (MoRTH, 2023). Intersections are one of the major pedestrian road crossing locations and they accounted for around 24% of total pedestrian deaths at intersections in the United States [National Highway Traffic Safety Administration (NHTSA), 2020 report]. This proportion may be higher in India where the traffic is heterogeneous and non-lane-based behavior is followed by the drivers. The high proportions of pedestrian deaths at intersections are because of the higher severity levels in P-V accidents because of the misunderstanding between vehicles and pedestrians due to dilemma behavior while crossing. A better understanding of P-V interactions is required to provide better pedestrian facilities at intersections to reduce safety-related issues.

In the past, researchers identified various factors affecting the severity levels of P-V interactions using accident data and SSMs. They also proposed the threshold limits for various severity levels of P-V interactions using either pedestrian or vehicle characteristics. But the severity depends on both pedestrian and vehicle characteristics. Determining the threshold limits for various P-V interaction severity levels using both pedestrian and vehicle characteristics will better correlate the actual severity in P-V interaction analysis and modeling.

The dilemma behavior of pedestrians while crossing is also one of the major reasons for P-V accidents. Misjudgment due to the dilemma behavior of pedestrians when the vehicle lies within a certain zone will cause P-V accidents. Providing the lower limit (unsafe gap) and upper limit (safe gap) for this dilemma behavior of pedestrians (known as PDZ) will improve the gap acceptance behavior and safety.

6.4 CHALLENGES IN IMPLEMENTATION, PARTICULARLY IN LOW- AND MIDDLE-INCOME COUNTRIES

Despite the potential benefits of surrogate safety analysis methods (SSAMs) for proactive road safety management, several challenges hinder their widespread implementation in low- and middle-income countries (LMICs). These challenges arise due to factors like heterogeneous traffic, lack of reliable data, limited technological infrastructure, and regulatory gaps.

6.4.1 Lack of Standardized Guidelines for SSAMs

Unlike the United States (Federal Highway Administration) or Europe, LMICs do not have official guidelines for using surrogate safety methods in traffic safety assessments. Indian Road Congress (IRC) standards and Ministry of Road Transport and Highways (MoRTH) road safety audits mainly rely on historical crash data, rather than predictive conflict-based safety evaluation. Absence of a national framework on SSAM limits its adoption in traffic engineering studies.

Some of the immediate actionable items for improving safety across the country are as follows:

- Development of Indian-specific SSAM guidelines under MoRTH and IRC.
- Mandatory inclusion of surrogate safety indicators (TTC, PET, Deceleration Rate to Avoid Crash) in road safety audits (similar to IRC: SP 88–2010) would help to overcome this problem.

6.4.2 Heterogeneous and Unorganized Traffic Conditions

- Unlike Western countries with lane-disciplined traffic, roads in LMICs have mixed traffic flow, including:
  - Motorcycles, auto-rickshaws, non-motorized vehicles, pedestrians, and street vendors sharing the same space.
  - Erratic driver behavior (lane violations, red-light jumping, sudden braking).
  - Frequent pedestrian crossings at uncontrolled locations.
- Existing microsimulation models (VISSIM, AIMSUN) struggle to accurately replicate Indian road behavior.
- Incorporation of real-world pedestrian movement patterns and two-wheeler behavior in traffic models.
- Adoption of deep-learning-based artificial intelligence (AI) conflict detection to identify chaotic traffic patterns.

6.4.3 Limited Availability of High-Quality Traffic Data and Trained Professionals

Surrogate safety methods require accurate traffic flow data collected through CCTV surveillance with AI integration. This will be a challenging task considering the enormous amount of infrastructure and budgetary requirements in LMICs. Most transportation engineers, urban planners, and policymakers in LMICs are not trained in SSAM methods. SSAM tools like VISSIM, AIMSUN, PTV Viswalk, and AI-based trajectory analysis require specialized knowledge that is not commonly taught in engineering curricula. Many advanced SSAM techniques require high-tech infrastructure, such as AI-powered video analytics (expensive high-resolution cameras, cloud-based processing), drones and LiDAR sensors for vehicle trajectory tracking, and advanced microsimulation software (VISSIM, AIMSUN, Simulation of Urban Mobility). Small and medium-sized cities cannot afford these technologies due to budget constraints.

6.4.4 Resistance from Traffic Authorities and Policymakers

Traditional crash-based safety evaluation is still the dominant method used by policymakers and traffic engineers. Lack of awareness about SSAM benefits results in resistance to adoption. Hence the road safety policies focus more on post-crash investigations rather than proactive conflict detection. Integration of SSAM conflict analysis into government-led black-spot identification programs paves way for a safer and better future.

6.5 SURROGATE SAFETY ANALYSIS USING EXTREME VALUE THEORY (EVT)

An important advantage of the extreme value theory (EVT) is that it will enable to model the stochastic behavior of unusually large or small processes in a given dataset. The key component of the EVT is to predict events that are rare, cross a threshold limit, and less frequent than previously observed events. Normally, this extreme behavior is rare and unobservable in the data collection period and this distinguished characteristic makes it applicable for surrogate safety prediction. Estimated crash probability based on specific crash proximity measures. Extreme sampling methods are frequently a source of serious concern.

These are critical events that do not result in a crash but are very close to that and can be used as SSMs (Tarko et al., 2009).

EVT is generally applied in terms of two approaches: one is block maxima (BM) approach and another is Peak over Threshold (POT) approach. These two approaches are generally applicable to estimate crash probabilities using surrogate safety indicators.

In the BM approach, observations are grouped into fixed intervals over time and space; a block of minima or maxima is treated as extremes. This shows the extreme values occurring at maximum over a period, such as an hour or a year. Whereas extreme event samples follow a generalized extreme value distribution, the samples are distributed in different blocks in a certain time interval, considering the largest and smallest threshold in each sample block. As BM approaches, R-largest order statistics are typically preferred as the majority of blocks have enough observations. BM approach mainly focuses on the behavior of independent random observation distributions.

> In the former case the method divides the sample time into blocks of a certain length and samples the largest value (or r largest values) in each block, whereas in the latter case all peak values are sampled and the values over a certain threshold are used to model the extremes.

POT as an extreme value if the related measurement exceeds the optimal predetermined threshold, where random observations $\{X_1, X_2, \ldots, X_n\}$ are identically and independently distributed within the Generalized Pareto Distributions (GPDs). For a suitable threshold, the GPD is as follows:

As for the BM approach each interaction can be considered as a block in which the minimum values of T2 and TTC are used. In the POT approach a threshold must be selected over which all the values are considered. The POT approach offers a different solution to modeling extreme events. It is necessary to choose a threshold over which extreme events are considered, in order that the theoretical justification applies thereby reducing bias.

However, the higher the threshold, the fewer available data remain. Thus, it is important to choose the threshold low enough to reduce the variance of the estimates.

There are basically two methods for selecting the appropriate threshold. (i) Mean residual plot: this plot shows the mean of the excesses depending on the value of the chosen threshold level u. Above a certain value, the GPD provides a valid approximation to the excess distribution. Here, a threshold has to be selected where the graph is linear within uncertainty bounds. This is, however, not always straightforward, and based on a subjective choice.

(ii) Model estimation: the model is estimated at a range of threshold values with the intention to find stable model parameters. Again, above a certain level of u the GPD is valid, if estimates of the shape parameter ξ are constant, while estimates of the scale parameter σ are linear in u. This point can be read from the plot by checking linearity; in other words, estimates will not change much within uncertainty bounds, as the threshold increases.

BIBLIOGRAPHY

Arun, V., Maurya, A. K., and Kharbeche, M. (2022). "Surrogate safety measures for evaluating heterogeneous traffic at urban intersections in India." *Transportation Research Part F: Traffic Psychology and Behaviour*, 82, 184–200.

Federal Highway Administration (FHWA). (2008). *Surrogate Safety Assessment Model (SSAM)*. U.S. Department of Transportation, (2025, November 14), https://highways.dot.gov/safety/zero-deaths

Gettman, D., Pu, L., Sayed, T., and Shelby, S. (2008). *Surrogate Safety Assessment Model and Validation: Final Report*. FHWA-HRT-08–051. U.S. Department of Transportation.

Hyden, C. (1987). "The development of a method for traffic safety evaluation: The Swedish Traffic Conflicts Technique." Doctoral dissertation, Lund University, Lund, Sweden. https://www.sciencedirect.com/topics/mathematics/extreme-value-theory

Indian Road Congress (IRC). (2010). Guidelines for Road Safety Audit (IRC: SP 88–2010). New Delhi, India.

Ministry of Road Transport and Highways (MoRTH), Government of India. (2023). Road Accident Data Report 2023. https://morth.nic.in

Mohamed, A., and Saunier, N. (2017). "Automated analysis of pedestrian-vehicle conflicts using video data and deep learning." *Transportation Research Record*, 2659(1), 55–64.

NHTSA (National Highway Traffic Safety Administration). (2020). *AI and Computer Vision for Traffic Safety Analysis*. U.S. Department of Transportation.

Patil, S., and Rathi, S. (2021). "Application of AI-based traffic video analytics for safety assessment in Indian urban roads." *Journal of Transportation Engineering, Part A: Systems*, 147(10), 04021055.

Pawar, D. S., Kumar, V., Singh, N., and Patil, G. R. (2016). "Analysis of dilemma zone for pedestrians at high-speed uncontrolled midblock crossing." *Transportation Research Part C: Emerging Technologies,* 70, 42–52.

Sharma1, S. N., Singh, D., and Dehalwar, K. (2024). "Surrogate safety analysis-leveraging advanced technologies for safer roads." *Suranaree Journal of Science and Technology*, 31(4), 14.

Tarko, A. P., Davis, G., Saunier, N., Sayed, T., and Washington, S. (2009). "Surrogate measures of safety in traffic engineering and their applications: Review." *Transportation Research Record: Journal of the Transportation Research Board*, 2103(1), 22–36.Song, Y., Tse, H., Lind, E. M., Chitturi1, M. V., and Noyce, D. A. Safety Evaluation of Transit Signal Priority with Bus Speed Volatility as a Surrogate Measure: Case Study in Minnesota. International Conference on Transportation and Development 2021: Transportation Operations, Technologies, and Safety held virtually June 8–10, 2021.

7 Simulation and Surrogate Safety

7.1 INTRODUCTION

"Surrogate safety" methodologies assess near-miss incidents or barely avoided collisions by utilizing trajectory data (from simulations or field recordings) alongside a defined set of formal metrics [e.g., Time-to-Collision (TTC), Post-Encroachment Time (PET), deceleration rates (DRs)]. Due to the infrequency and randomness of crashes, surrogate measures enable analysts to assess safety performance and evaluate alternatives (designs, signal timing, Connected and Automated Vehicle (CAVs)) proactively, without the necessity of accumulating crash histories. The Surrogate Safety Assessment Model (SSAM) and prior Federal Highway Administration (FHWA) research established methods, suggested thresholds, and procedures for deriving surrogate safety measures (SSMs) from trajectory information.

7.2 SIMULATION APPLICATIONS

Traffic microsimulation + automated conflict analysis is used in at least the following applications:

1. **Design alternatives and intersection/roundabout safety** quantify expected changes in conflict counts/severities when geometry or control is changed (gap acceptance at unsignalized intersections, turn pocket lengths, roundabout yield layouts).
2. **Signal timing/coordination tradeoffs** evaluate safety-performance tradeoffs (e.g., shorter cycle splits reduce queue but may increase rear-end conflicts). Microsimulation + SSMs are standard for this.
3. **New technologies (Advanced Driver Assistance Systems ADAS/CAV/ connected vehicle strategies)** evaluate various penetration rates, warning/actuation algorithms, and lane management strategies, and quantify surrogate safety enhancements or decline before to field implementation (Figure 7.1).
4. **Proactive hotspot screening:** short-duration trajectory samples (from Connected Vehicles (CVs) data or video drone captures) feed SSM models to find high-risk segments faster than waiting for crash accumulation.
5. **Scenario generation and validation datasets** generate representative pre-crash scenarios for AD/ADAS testing or human factor evaluation by sampling from SSM-selected events (Figure 7.2).

DOI: 10.1201/9781003604761-7

FIGURE 7.1 Traffic simulation model and geometry

Source: https://blog.ptvgroup.com/en/technologyplus/realistic-traffic-simulation-driving-behavior-is-key/

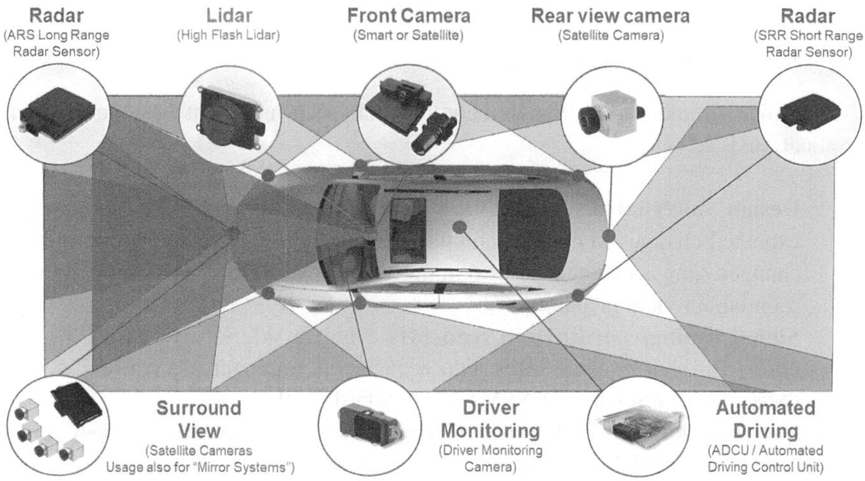

FIGURE 7.2 Advanced Driver Assistance Systems (ADAS) system in vehicles

Source: https://www.spinny.com/blog/adas-full-form/

How the workflow normally runs (high level): Develop a calibrated microsimulation or utilize field trajectories, export trajectory files (vehicle time-stamped locations, velocities), and execute an automated SSM tool (SSAM or alternative) to identify conflict situations and calculate SSMs. Post-process results (counts, exposure rates, severity distribution) lead to statistical modeling or ranking and decision support.

7.3 SIMULATION—KEY PARAMETERS (WHAT TO SET, WHY THEY MATTER)

A safe, defensible surrogate analysis requires careful attention to both model parameters and SSM configuration. The items below are the ones with the largest effect on SSM outputs.

Driver-Behavior and Car-Following/Lane-Change Models

- **Car-following**: model family [psychophysical, Wiedemann used in VISSIM; IDM (Treiber), continuous family; Krauss used in SUMO] controls spacing, reaction, and accelerations; these directly change TTC distributions and lane-change behaviors. Always find which model the simulator uses and its parameter set (Figure 7.3).
- **Lane-change logic**: thresholding for gap acceptance, politeness, and anticipation strongly alter lateral conflicts; tune using trajectory data. (Different lane-change models produce very different lateral conflict counts.)

Human Response/Perception Parameters

- Perception–reaction time, aggressiveness (desired headway), gap acceptance, and critical deceleration are central to conflict severity (they change whether a vehicle brakes or performs evasive action). These must be calibrated to observed driver behavior for the study population.

Vehicle Dynamics

- Maximum accel/decel, vehicle lengths, and truck/car dynamics influence distance/speed at the conflict point and hence delta-V (ΔV) and deceleration SSMs. Mixed-fleet parameterization (cars vs. heavy vehicles) is important where truck shares are non-negligible.

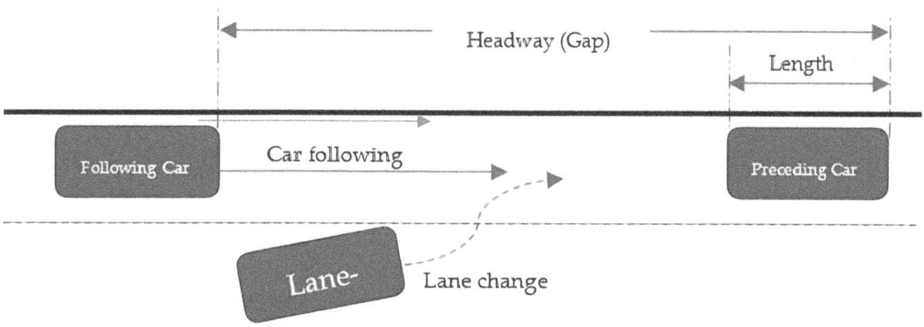

FIGURE 7.3 Lane-changing behavior

Source: https://www.researchgate.net/figure/Typical-illustration-for-car-following-lane-change-and-gap-acceptance_fig1_350130840

Environment and Demand Parameters

- Traffic volumes, route choice, vehicle mix, turning percentages, pedestrian/bike flows, visibility/lighting, and weather assumptions all affect SSM counts. In particular, demand relative to capacity (degree of saturation) is a first-order driver of conflict frequency.

Simulation Control and Statistical Settings

- The time step (simulation resolution), simulation warm-up, and quantity of stochastic replications (varied random seeds) are essential. Due to the stochastic nature of microstimulators, it is advisable to conduct several replications and calculate confidence intervals for SSM counts. Agency advice outlines techniques for determining the necessary number of replications to achieve a specified confidence interval. Standard procedure involves 10–30 replications for operational studies; calculate the precise N based on the desired confidence interval and sample variation.

SSM Detection Thresholds and Classification

- The TTC/PET thresholds and conflict angles define the parameters of an event. SSAM defaults are frequently employed as initial parameters (TTC = 1.5 s, PET ≈ 5 s); however, the selection of thresholds necessitates justification and sensitivity analysis, as it influences counts and correlations with crashes. Algorithms categorize conflict types (rear-end, lane-change, path-crossing) based on approach angles.

Output SSMs (Computed Measures)
Common SSMs calculated by SSAM and other tools include:

- Minimum TTC: the time until collision if velocities remain constant. (See Section 7.4 for formula.)
- PET: time between when one road user leaves the conflict point and the other arrives (PET = $t_2 - t_1$), useful for crossing conflicts.
- Initial DR, maximum deceleration (MaxD), maximum speed (MaxS), ΔV (velocity change if collision occurred).

Implementation Recommendations (Short Checklist)

- Record trajectory data with adequate temporal resolution (0.1 s or better when possible).
- Calibrate first at macroscopic (flows, speeds) and microscopic targets (headway distribution, lane-change rates, trajectory shapes). Use trajectory databases where possible (NGSIM/highD/INTERACTION) to calibrate micro-parameters.
- Always run sensitivity analysis on TTC/PET thresholds and number of replications (random seeds).

7.4 CORE SURROGATE MEASURES FOR CONFLICT ANALYSIS

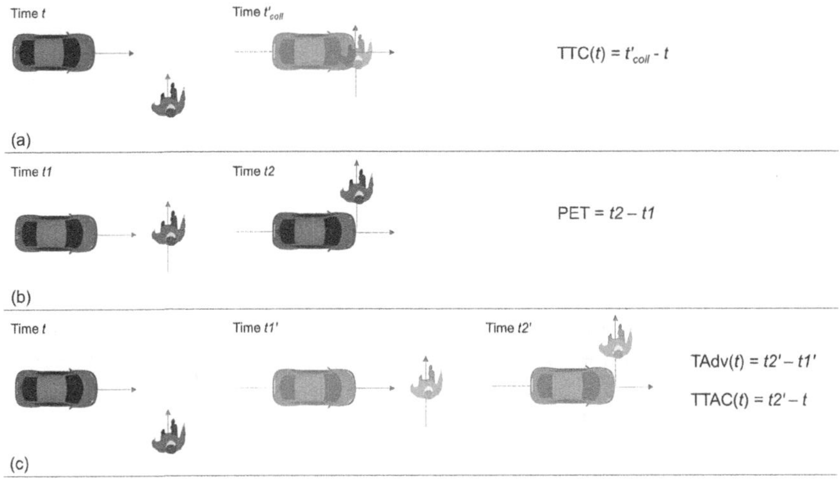

FIGURE 7.4 (a) time-to-collision (TTC), (b) post-encroachment time (PET), (c) time advantage (TAdv) and time-to-avoided-collision-point (TTAC). Solid figures represent noticed scenarios, while transparent figures denote anticipated scenarios.

Source: https://www.researchgate.net/figure/sual-representation-of-a-time-to-collision-TTC-b-post-encroachment-time-PET_fig1_372075005

7.4.1 Time-to-Collision

TTC is the predicted time remaining until two road users collide if they maintain their current (or modeled) motions. It's most informative for rear-end and same-path situations (approaching a stopped queue, platoon disturbances, ramp merges on the same lane). The concept goes back to Hayward's "near-miss" scale of danger (Figure 7.4).

Let the lead vehicle be i and the following vehicle be j. Use front-to-rear gap $g_0 > 0$ at time $t = 0$.

- **Constant-speed TTC:**

$$\text{TTC} = \frac{g_0}{v_j - v_i}, v_j > v_i \tag{7.1}$$

$$\text{TTC} = \infty, v_j \leq v_i \tag{7.2}$$

(Only defined when the follower is closing.)

- **Constant-acceleration TTC (more realistic when braking occurs)**: Solve for the first positive root of

$$d(t) = g_0 + (v_i - v_j)t + \frac{1}{2}(a_i - a_j)t^2 = 0 \tag{7.3}$$

With $\Delta v = v_i - v_j$ and $\Delta a = a_i - a_j$:

$$t = \frac{-\Delta v \pm \sqrt{\Delta v^2 - 2\Delta a g_0}}{\Delta a}, \tag{7.4}$$

choosing the smallest $t > 0$ if the discriminant is ≥ 0. If $\Delta a = 0$ this reduces to the constant-speed case.

How it's used in tools? In FHWA's SSAM, analysts often classify a "conflict" when minimum TTC during an interaction is ≤ 1.5 s (default, adjustable).

Interpreting TTC

- Low TTC ⇒ high immediacy of risk (likelihood of crash without evasive action), but not injury severity (speed isn't directly in TTC). FHWA's early SSAM work cautions on this point.
- Always examine TTC distributions (percentiles) and minimum TTC over an encounter, not single frames.

Example A (freeway following with sudden braking): Lead car at 20 m/s (~72 km/h) begins braking at –3.0 m/s². Follower at 22 m/s (no braking yet). Initial front-to-rear gap g_0 = 25 m.

- Constant-speed TTC would be $\frac{25}{22-20} = 12.5$ s (misleadingly "safe").
- Constant-acceleration TTC: $\Delta v = 20 - 22 = -2$ m/s; $\Delta a = -3 - 0 = -3$ m/s^2 gives

$$t = \frac{-(-2) - \sqrt{(-2)^2 - 2(-3)(25)}}{-3} = \frac{2 - \sqrt{4 + 150}}{-3} \approx 3.47 \text{ s.}$$

The physically correct (positive) root is ≈ 3.5 s, revealing a near-term hazard the constant-speed TTC misses.

Example B (ramp merge, same lane): A merging car accelerates from 16 m/s to 22 m/s, while the through vehicle maintains a speed of 25 m/s, with a separation of 30 m between them. Initially, $v_j < v_i$ implies TTC = ∞. However, after v_j surpasses v_i, the TTC becomes finite; calculating the minimum TTC within the merging window indicates whether a lane change is advisable or if metering is required.

Implementation Tips and Pitfalls

- **Sampling and smoothing**: Use ≥ 10 Hz trajectories; smooth speeds/accelerations (e.g., Savitzky–Golay) to avoid TTC flicker from noise.
- **Vehicle length and gap**: Always measure front-to-rear gap, not centroid distance, or you'll bias TTC high.
- **Sign conventions**: Guard against negative/complex roots (no collision under assumed motion).
- **Mode suitability**: TTC is strongest on same-path conflicts; for path-crossing, PET is usually more telling.

7.4.2 Post-Encroachment Time

PET is the time between one road user leaving a conflict area and the other user arriving at that same area. It shines at intersections, crosswalks, and roundabouts, anywhere paths cross, but a collision is barely avoided. PET was introduced by Allen, Shin, and Cooper.

Let t_{leave} be the instant the first user exits the conflict area (the overlapping space they both would occupy), and t_{arrive} be the time when the other user's reference point enters it:

$$\text{PET} = t_{\text{arrive}} - t_{\text{leave}}\,,\ \text{PET} \geq 0. \tag{7.5}$$

A PET of 0 means they reached the area at the same time (a modeled crash).

Typical thresholds: SSAM's widely used default flag conflicts when PET ≤ 5.0 s (and TTC ≤ 1.5 s, jointly). For pedestrians/cyclists, researchers often adopt larger PET cutoffs (e.g., 7–9 s) due to vulnerability and differences in motion.

How to Compute It Robustly (from Video or Simulation)

1. Define the conflict area (polygon) from the two paths' overlap (lane lines, crosswalk, turning path).
2. Track each user's reference point or swept rectangle (using actual dimensions for precision).
3. Find t_{leave}: Last moment user A intersects the conflict area.
4. Find t_{arrive}: First moment user B intersects it after t_{leave}.
5. Interpolate to sub-frame time using linear (or constant-accel) interpolation of position.

Resolution note: With 30 fps video, your PET resolution is ≈ 0.033 s; with 10 Hz simulation output, it's 0.1 s; interpolation matters.

Example C (unsignalized crosswalk, car vs. pedestrian): A pedestrian clears the last 0.5 m of the crosswalk at $t = 12.40$ s. The oncoming car's front bumper reaches the crosswalk at $t = 14.95$ s.

$$\text{PET} = 14.95 - 12.40 = 2.55\,\text{s}.$$

A short miss for a vulnerable road user (VRU). Many studies would classify this as a serious conflict for VRUs (consider PET thresholds ≥ 7–9 s).

Example D (signalized left-turn vs. opposing through): A left-turning SUV exits the conflict area at $t = 36.2$ s; the opposing sedan arrives at $t = 39.1$ s. PET = 2.9 s. If the site's accepted PET threshold is 5 s (SSAM default), then this is a moderate conflict; if crash history shows angle crashes, then you'd consider protected phasing or change the permissive window.

Strengths and Pitfalls

- **Pros**: Captures how close in time two users miss each other; great for path-crossing.

- **Cons**: Needs a well-defined conflict polygon; occlusions can bias timestamps; a large PET isn't automatically safe if speeds are very high just after the conflict point.
- **Mode sensitivity**: Choose thresholds per mode; SSAM vehicle-vehicle defaults don't necessarily fit VRUs.

7.4.3 DELTA-V

ΔV is the vector change in a vehicle's velocity due to impact, the integral of crash acceleration over the pulse. It is a standard crash-severity descriptor used by NHTSA/IIHS and forms the backbone of reconstruction tools (e.g., CRASH/SMASH families). Higher ΔV generally correlates with greater injury risk, though pulse duration, direction, restraint use, and intrusion also matter.

For vehicle k:

$$\Delta V_k = V_{k\text{ post}} - V_{k\text{ pre,}} \quad |\Delta V_k| = \text{ Severity magnitude.} \tag{7.6}$$

Compute using momentum and a coefficient of restitution "e" along the line of impact (Figure 7.5).

- **1D, same-direction (rear-end) closed-form** (masses m_1, m_2; pre-impact speeds u_1, u_2; restitution e):

$$v_1 = \frac{m_1u_1 + m_2u_2 + m_2e(u_2 - u_1)}{m_1 + m_2}, \quad v_2 = \frac{m_1u_1 + m_2u_2 - m_1e(u_2 - u_1)}{m_1 + m_2}. \tag{7.7}$$

Then $\Delta V_1 = v_1 - u_1$, $\Delta V_2 = v_2 - u_2$. (7.8)

- **2D, right-angle (e.g., intersection) with $e \approx 0$:**
  Use vector momentum: $V_{\text{post}}=(m_Au_A + m_Bu_B)/(m_A + m_B)$.
  Each car's ΔV is $V_{\text{post}} - u_k$.

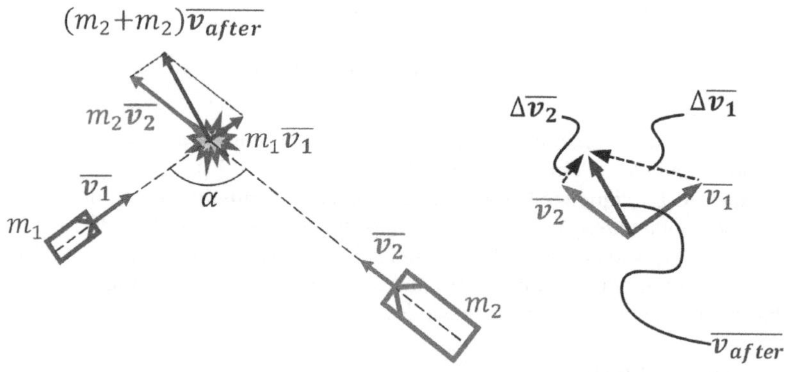

FIGURE 7.5 Delta-V

Source: https://ars.els-cdn.com/content/image/1-s2.0-S0001457516303566-gr3.jpg

Field note: Modern reconstructions often take ΔV directly from event data recorders (EDRs) when available; otherwise from crush and momentum methods.

Example E (rear-end, equal masses, light restitution): Lead car (1): m_1 = 1,500 kg at u_1 = 10 m/s (36 km/h); follower (2): m_2 = 1,500 kg at u_2 = 18 m/s (64.8 km/h); e = 0.2.

$$v_1 \approx 14.8 \text{ m/s}, \quad v_2 \approx 13.2 \text{ m/s}.$$

ΔV_1 = +4.8 m/s, ΔV_2 = −4.8 m/s (magnitudes ≈ 4.8 m/s = 17.3 km/h).

This is a moderate ΔV rear-end; injury risk rises with ΔV, but true occupant risk also depends on belt use and pulse duration.

Example F (urban right-angle, inelastic): Car A: 1,500 kg eastbound at 50 km/h (13.9 m/s). Car B: 1,200 kg northbound at 40 km/h (11.1 m/s). Assume they lock up (e = 0) at the intersection.

Post-impact velocity of the combined mass:

$$V_{\text{post}} \approx (8.0, 4.8) \text{ m/s}.$$

Magnitudes of ΔV:

- Car A: $|\Delta V_A| \approx$ 7.9 m/s (28.5 km/h).
- Car B: $|\Delta V_B| \approx$ 9.9 m/s (35.6 km/h).

Right-angle impacts typically generate larger ΔV for a given approach speed than gentle rear-ends—one reason their injury outcomes are often worse. (Direction matters too: lateral ΔV is more injurious for unprotected occupants.)

How to Use ΔV in Proactive (Surrogate) Analysis

- From simulations, you can approximate a *predicted* ΔV by integrating modeled crash pulses where TTC → 0 or PET → 0.
- In conflict-only (no actual crash), use extended ΔV proxies (e.g., "ΔV" if no evasive action after $t = t^*$) or complementary severity measures [Deceleration Rate to Avoid Crash (DRAC), speed at conflict point]. See review work on extended/derived measures.

Strengths and Pitfalls

- **Pros**: Physically grounded, widely used in crash analysis and injury research.
- **Cons**: True ΔV exists only if a crash occurs; predicted ΔV depends on your collision model (contact direction, restitution, crush, EDR availability).

Choosing among TTC, PET, and ΔV (Quick Guidance)

- **Rear-end and same-lane**: TTC (plus DRAC) for likelihood; estimate ΔV for severity if collision becomes unavoidable.
- **Intersections and crosswalks**: PET is usually the clearest *temporal miss* metric; pair with speeds at the conflict point to reflect severity. SSAM

defaults of TTC ≤1.5 s and PET ≤5 s are common for vehicle-vehicle; raise PET thresholds for VRUs.

- **Severity emphasis**: ΔV (actual or predicted) relates best to injury risk but requires crash modeling/EDR.

Common Implementation Pitfalls (and How to Avoid Them)

1. **Noisy trajectories → spurious extremes**: Smooth positions before differentiating; constrain accelerations to plausible limits (e.g., ±5 m/s^2 for passenger cars in steady dry conditions).
2. **Wrong geometry for PET**: Use swept vehicles (or buffered polylines), not point-centroids, to define the conflict polygon.
3. **Frame-rate bias**: Interpolate to sub-frame times for TTC/PET minima.
4. **One-size thresholds**: Document your chosen TTC/PET cutoffs by mode and context; SSAM defaults are a starting point, not a standard.
5. **TTC ≠ severity**: Combine with speed at conflict, conflict angle, or predicted ΔV when you need severity ordering.

7.5 RELEVANCE WITH FIELD PARAMETERS (MAPPING SIMULATION → REALITY)

The central practical question for an analyst is: how representative are SSMs of real-world risk simulated? Below are the key steps, empirical links, and pitfalls.

Data Sources (Field Inputs and Validation Targets)

- **Trajectory datasets** (for calibration and validation): NGSIM (detailed trajectories of US motorways and arterials), highD (drone trajectories on German highways), INTERACTION, inD, and rounD (for junction and roundabout interactions) offer the microscopic benchmarks necessary for calibrating car-following and lane-change algorithms.
- **Naturalistic driving and pre-crash datasets (NDS/SHRP2)** useful for behavior and pre-crash kinematics for crash generation/validation (driver response, distraction prevalence).
- **Connected vehicle Basic Safety Messages** are a growing source of large-scale trajectory-like samples from fleets, promising for near-real-time surrogate analytics when market penetration is sufficient.

Calibration Targets and Methods

- **Macro targets**: link flows, lane flows, speeds, travel times.
- **Micro targets**: headway distribution percentiles, time-to-lane-change, lane-change rates, observed TTC/PET distributions. Trajectory-level calibration is increasingly recommended (calibrating on trajectories rather than only aggregate outputs).

Exposure and Normalization

- Express SSM outputs as rates to compare locations or alternatives (e.g., conflicts per million vehicle-km, conflicts per 1,000 vehicle-hours, or conflicts per 10,000 turning movements). Rate denominators should align with the policy question (volume-based for network management; movement-based for intersection treatments).

From Surrogate Counts to Expected Crashes—Empirical Calibration

- Numerous studies find significant correlations between SSM counts and historical crash counts, but the relationship is imperfect and site/context dependent; therefore, empirical calibration (statistical models that map SSM counts and traffic covariates to observed crashes) is required before using SSMs as direct crash predictors. SSAM validation work shows that SSMs correlate with crash frequencies for many intersection types, but the strength of correlation varies by geometry and volume.

7.6 RISK ESTIMATION—TURNING SSMS INTO PROBABILISTIC/FREQUENCY MEASURES

This section provides practical modeling routes (with pros/cons) for estimating risk from SSM outputs.

Two Common Objectives and How They Differ

1. **Frequency prediction**: estimate the expected number of crashes (counts) per period at a site using SSM counts as predictors [useful for hotspot ranking and empirical Bayes (EB) adjustments].
2. **Episode-level probability**: estimate the probability that a given conflict event (or driving episode) would have resulted in a crash (useful for on-vehicle warning algorithms and scenario generation).

7.6.1 Statistical Modeling Approaches

A. Crash-frequency models augmented with SSMs

- Negative binomial (NB) or Poisson models are conventional for crash counts; using SSM measures (conflict rates, mean minimum TTC, counts below a threshold) enhances explanatory power in numerous studies. Employ model diagnostics (Akaike Information Criterion, marginal R^2) and cross-validation. Recent research indicates that using both proximity (TTC) and avoidance (DR) SSMs can significantly enhance NB models.

B. EB before-and-after frameworks using SSMs

- EB functions as the standard for calibrating regression-to-the-mean effects and variations in traffic during safety evaluations. SSMs can function as ancillary data in EB priors or to develop jurisdictional Safety Performance Functions (SPFs) that integrate SSM predictors. FHWA workshops and texts delineate the incorporation of many predictors into EB.

C. Episode-level logistic/survival models

- Use logistic regression (or survival/transition models) on labeled episodes (from SHRP2/NGSIM where near-crash/crash labels exist) to map TTC/PET/ΔV to crash probability. When the data are imbalanced (crashes are rare), use stratified sampling, case-control adjustments, or extreme-value approaches.

D. Bayesian and hierarchical models

- It is useful to pool information across sites and account for random effects (site heterogeneity), and to propagate parameter uncertainty to risk estimates. For conflict-based before-and-after studies, full-Bayes approaches can be more robust than simple EB under certain conditions.

E. Machine learning

- Tree-based models (such as random forests and XGBoost) have been employed with SSMs and geometric/traffic data for hotspot categorization, demonstrating significant predictive efficacy; nonetheless, they necessitate meticulous regularization and elucidation for policy implementation. Consistently juxtapose machine learning outcomes with traditional statistical methods for enhanced interpretability and uncertainty assessment.

Practical Step-by-Step Recipe to Estimate Expected Crashes Using SSMs

1. Compute SSMs for your baseline period in simulation or field samples, and express exposure-normalized rates (e.g., conflicts per 1,000 vehicle-hours).
2. Gather historical crash counts and traffic covariates for comparable sites (if you plan to model frequency).
3. Fit a crash-frequency model (Poisson or NB) where SSM rates are candidate predictors; consider transformations (log, percentile counts) and interaction terms. Use hold-out cross-validation.
4. If doing before-and-after evaluation, use EB to combine site record with SPF predictions (SPF can include SSM covariates), then estimate expected crash reduction and confidence intervals.
5. Report both point estimates and uncertainty (95% CI); perform sensitivity to TTC/PET threshold choice and the number of simulation replications.

7.7 VALIDATION AND LIMITATIONS (DETAILED)

Validation Strategies (Recommended Hierarchy)

1. **Tool-level check**: confirm whether the SSM tool computes the same measures for a small hand-calculated scenario (sanity checks for TTC/PET).
2. **Trajectory-level validation**: compare simulated and field trajectory statistics (speeds, headways, lane changes). Use NGSIM/highD/INTERACTION datasets where possible for calibration/validation.
3. **Conflict detection validation**: manually review a random sample of conflicts from recorded video (or simulated replay) and check that the automated tool's event list matches human coders' judgments (compute confusion matrix, precision, recall). Recent studies show video-tool concordance varies by lighting/occlusion and conflict type.
4. **Aggregate external validation**: estimate correlation/regression between SSM rates and historical crash counts across many sites; significant positive correlations support the use of SSMs for screening, but do not guarantee transferability.
5. **Predictive validation**: if possible, withhold some sites/periods and test whether SSM-augmented models can predict crash counts or future high-risk sites (out-of-sample). Compare Receiver Operating Characteristic Area Under the Curve (ROC/AUC), Root Mean Square Error (RMSE), and rank correlations (Spearman) depending on the objective.

Limitations and Failure Modes (Be Explicit)

- **Threshold sensitivity**: TTC/PET threshold choice is arbitrary to an extent and affects counts dramatically; analysts must run sensitivity analysis and, where possible, derive thresholds calibrated to observed crash risk. Default SSAM values (TTC ≈ 1.5 s, PET ≈ 5 s) are a useful starting point but not universal.
- **Behavioral realism**: microsimulation models are approximations of human behavior. If driver models are mis-calibrated (e.g., aggressive real drivers vs. conservative model parameters), SSM counts will be biased. Trajectory-level calibration mitigates but does not fully remove this.
- **Representativeness and exposure**: a short sample of trajectories (video or CV) may not capture seasonal/temporal variations; SSM rates must be standardized to relevant exposure metrics and sampling biases addressed.
- **Data quality (occlusion and tracking errors)**: video and sensor extraction produce errors (false tracks, lost tracks) that can create false conflicts or miss real ones. Drone/aerial data largely reduces occlusion but has its processing errors. Validate trackers against manual coding when possible.
- **Transferability**: relationships (SSM → crash) are often locality-specific (driver culture, road geometry, enforcement). Avoid directly porting a

calibrated conversion function from one jurisdiction to another without testing.

- **Rare events and severity mapping**: crashes are rare and often result from a confluence of factors; SSMs capture proximity and evasion but may under-represent non-conflict crashes (single-vehicle run-off, road object impacts) and some severe crash precursors. Complement SSM analyses with other measures when evaluating total risk.

Practical Mitigation (What to Do)

- Document algorithms, thresholds, and model versions for reproducibility.
- Calibrate micro-parameters to trajectories (not just flows); use multiple datasets if available.
- Run uncertainty analysis (replications, bootstrap of SSM counts, sensitivity to TTC/PET). Report CIs.
- Combine SSMs with crash-based methods (EB/SPF) rather than replacing crash-based inference entirely.

7.7.1 Short Worked Illustration (Conceptual)

Suppose you have a pair of vehicles on the same lane with the following instantaneous values at time t:

- Lead vehicle position $x_i = 100.0$ m, length $L_i = 4.5$ m, speed $v_i = 20.0$ m/s.
- Following vehicle position $x_j = 80.0$ m, speed $v_j = 25.0$ m/s.

Relative bumper-to-bumper distance $= x_i - x_j - L_i = 100.0 - 80.0 - 4.5 = 15.5$ m.

Relative speed (closing) $= v_j - v_i = 25.0 - 20.0 = 5.0$ m/s.

TTC = 15.5/5.0 = 3.10 s. (If the analyst's TTC threshold is 1.5 s, this would not be recorded as a critical conflict; if the threshold was 4.0 s, it would.) This example shows how both spacing and closing speed combine non-linearly to produce TTC, and why threshold choice matters. (Numbers are illustrative.)

Practical Checklist for an SSM-Based Safety Study

1. Define objective (screening, before-and-after, design comparison, ADAS testing).
2. Choose/collect trajectories (simulation or field). Ensure temporal resolution ≥ 0.1 s when possible.
3. Calibrate microsimulation at macro and micro levels (flows, speeds, headways). Document parameter values.
4. Decide SSM thresholds and justify them (use defaults only after sensitivity testing).
5. Run sufficient stochastic replications and compute confidence intervals.

6. Validate: tool-level checks, trajectory comparisons, manual conflict validation sample, and external aggregate tests vs. historical crashes.
7. If producing crash estimates, use EB or NB regression with SSM predictors and report uncertainty.

BIBLIOGRAPHY

Allen, B. L., Shin, B. T., and Cooper, P. J. (1978). Analysis of Traffic Conflicts and Collisions. TRR 667. (PET origin/definition).

FHWA -HRT-08–051, SSAM and Validation—Final Report (2008).

FHWA Tech Briefs and docs noting TTC = 1.5 s and PET = 5.0 s defaults.

Gettman, D., and Head, L. (2003). Surrogate Safety Measures from Traffic Simulation Models, FHWA-RD-03–050, Federal Highway Administration. (SSM definitions, surrogate measures, early FHWA synthesis).

Hauer, E. (1997). Observational Before–After Studies in Road Safety, Pergamon. See also: FHWA Tutorial Estimating Safety by the Empirical Bayes Method: A Tutorial. (EB method for safety evaluation).

Hayward, J. C. (1972). Near-Miss Determination Through Use of a Scale of Danger. Highway Research Record 384. (TTC origin/definition).

Hussein, S., et al. (2008). Surrogate Safety Assessment Model and Validation: Final Report, FHWA-HRT-08–051, Federal Highway Administration (SSAM final report; algorithms, validation, recommended defaults).

Krajewski, R., Bock, J., Klöcker, L., and Eckstein, L. (2018). The highD Dataset: A Drone Dataset of Naturalistic Vehicle Trajectories on German Highways, IEEE/ITSC (dataset and description). (Large naturalistic trajectory dataset).

Krauss, S. (and SUMO documentation). Krauss car-following model (SUMO docs) open-source car-following family used in SUMO.

Pedestrian/Vulnerable road user threshold adaptation, Example calibration showing larger PET/TTC cutoffs for VRUs.

Recent review & empirical work: "A review of surrogate safety measures" and "Enhanced Crash Frequency Models Using Surrogate Safety Measures" (2023) for advances that combine connected vehicle data and modern statistical models; see Liu et al. (2023)/ TRID/SAGE for examples.

SSAM Software User Manual (FHWA). Surrogate Safety Assessment Model (SSAM) – Software User's Guide (FHWA-HRT-10–020/TechBrief). (Software operational guidance; threshold defaults and outputs).

Treiber, M., Hennecke, A., and Helbing, D. (2000). "Congested traffic states in empirical observations and microscopic simulations." *Physical Review E*, 62, 1805–1824. (IDM car-following model).

NGSIM, 2005.NGSIM US 101 Data Analysis. Next Generation Simulation (NGSIM) vehicle trajectories and supporting data (public dataset). (Trajectory benchmark often used to calibrate micro-models), Federal Highway Administration, U.S. DOT/FHWA https://www.kaggle.com/datasets/nigelwilliams/ngsim-vehicle-trajectory-data-us-101

Validation and applied studies: Validation of SSAM for urban intersections, and more recent drone/vision-based validation work (examples cited in text).

Wiedemann, R. (1974, 1999). Psychophysical car-following model (Wiedemann models) — see model summaries used in VISSIM, calibration studies, and modern analyses.

ΔV and crash severity, NHTSA primers on ΔV and crash pulse interpretation; IIHS backgrounders (ΔV as severity descriptor).

8 Pedestrian Surrogate Safety Procedures

8.1 INTRODUCTION

Pedestrians are among the most vulnerable road users, as they lack the physical protection that vehicle occupants have. Pedestrian-vehicle (P-V) interactions occur in various environments, including crosswalks, sidewalks, intersections, and shared spaces, making pedestrian safety a critical concern. According to the World Health Organization (WHO), pedestrians account for approximately 23% of global road traffic fatalities. The increasing urbanization, high vehicle volumes, and non-uniform pedestrian movement patterns contribute to a high number of P-V conflicts, often resulting in severe injuries or fatalities.

In India, pedestrian safety is a major concern due to rapid urbanization, increasing motorization, and insufficient pedestrian-friendly infrastructure. According to the Ministry of Road Transport and Highways (MoRTH) 2023 report, pedestrians account for 17% of total road fatalities in India. The key issues contributing to this high fatality rate include:

- Poor enforcement of pedestrian right-of-way laws.
- Lack of pedestrian-exclusive zones in high-footfall areas.
- Ineffective traffic calming measures near pedestrian crossings.

Pedestrian safety is vital for promoting sustainable urban mobility, reducing road fatalities, and encouraging walking as a primary mode of transportation. Safe pedestrian infrastructure benefits all road users by:

- Reducing P-V conflicts.
- Enhancing accessibility for children, the elderly, and people with disabilities.
- Improving urban livability by encouraging non-motorized transport.

8.2 PEDESTRIAN-VEHICLE INTERACTIONS

Several factors contribute to P-V conflicts and accidents, including:

Road and Traffic Characteristics

i. High-speed roads: Pedestrian safety is significantly reduced on roads with high vehicle speeds, especially where crossings are absent or poorly designed.

DOI: 10.1201/9781003604761-8

ii. Unsignalized intersections: Many P-V conflicts occur at uncontrolled intersections where no signals or traffic control devices exist to regulate movement.
iii. Mixed traffic conditions: In countries like India, roads are shared by various users, including motorized vehicles, cyclists, and pedestrians, leading to unpredictable interactions.

Infrastructure Deficiencies

i. Lack of pedestrian crossings: Many roads lack designated pedestrian crossings, forcing people to cross at unsafe locations.
ii. Poor sidewalk design: Narrow or obstructed sidewalks discourage pedestrians from using designated pathways.
iii. Inadequate lighting: Poorly lit roads contribute to pedestrian crashes, particularly at night.

Human Factors

i. Pedestrian behavior: Risky behaviors such as jaywalking, crossing mid-block, or using mobile phones while walking increase pedestrian crash risks.
ii. Driver non-compliance: Many drivers do not yield to pedestrians at crosswalks, even where laws mandate it.
iii. Lack of awareness: Both pedestrians and drivers may be unaware of traffic rules related to pedestrian safety.

8.3 PEDESTRIAN CRASH RISK ASSESSMENT

8.3.1 Pedestrian Dilemma Zone

When a pedestrian is crossing the road, he/she may try to find a sufficient gap and the decision to stop/go depends on the available gap size. But sometimes pedestrians may get confused about whether the available gap is sufficient to cross the road or not. This situation is known as the dilemma of a pedestrian. This situation arises only when the vehicles are present within a certain section of a roadway and this section is known as the pedestrian dilemma zone (PDZ). Gap acceptance behavior is the basic concept for dilemma zone analysis. PDZ is defined as "the section of a roadway where the presence of vehicles creates a stage of confusion for pedestrians while crossing the road" (Pawar et al., 2016). It consists of upper and lower boundary limits and the difference between these two limits is defined as the length of the dilemma zone. Figure 8.1 shows the physical location of a PDZ for one approach near an intersection. D_0 and D_1 are the lower and upper boundary limits respectively within which pedestrians are in a stage of confusion to judge whether the available gap is sufficient or not. There are several methods to estimate the dilemma zone using the filed data and the basic procedures of various methods to estimate PDZ boundary limits were briefly discussed below.

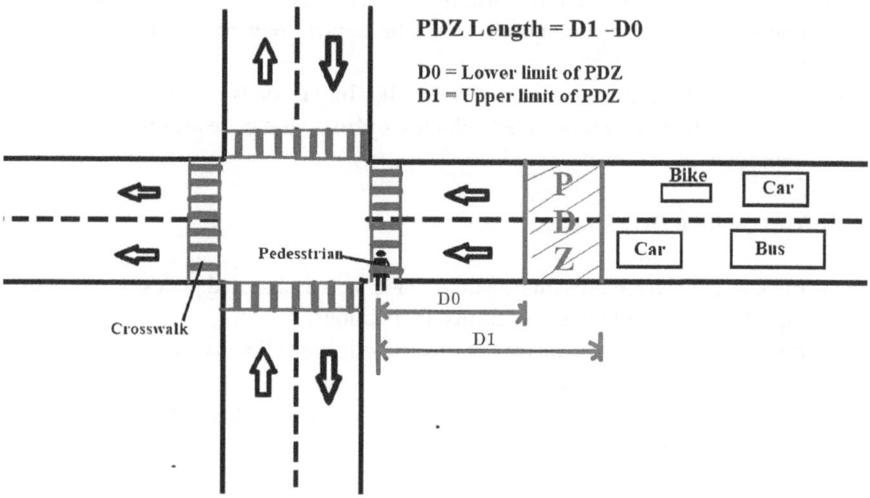

FIGURE 8.1 Location and length of pedestrian dilemma zone at an intersection

i) **Gap cumulative distribution method**: This method is proposed to estimate the dilemma zone boundary limits. In this method, a plot between the cumulative percentage gap accepted/rejected on y-axis and distance on x-axis is plotted and then the distance on x-axis corresponding to 10 percentile accepted and the 90 percentile rejected gaps on y-axis are taken as the lower and upper boundary limits of dilemma zone respectively.

ii) **Support vector machine (SVM) method**: **Pawar et al. (2016)** used the SVM technique to find the dilemma zone boundaries at an uncontrolled midblock crossing. In this method, a plot between the speeds of vehicles or pedestrians on y-axis and distances of the vehicles from the pedestrian/vehicle trajectory paths on x-axis is plotted on a coordinated system. A hyperplane is constructed as a decision plan to separate the accepted and rejected gap sizes with maximum margin. The lower and upper limits of the dilemma zone are taken on the hyperplane corresponding to the plus or minus two standard deviations from the mean speed.

iii) **Binary logit (BL) method**: This method has two alternative outputs (accept/reject) as it is a discrete choice model. In this method, a BL model is developed using gap accept/reject as a dependent variable and various independent variables, and the lower and upper boundary limits of dilemma zone are estimated corresponding to the 0.1 and 0.9 probabilities.

iv) **Probabilistic method**: This method was proposed by Gazis et al. (1960) for estimating the dilemma zone boundaries. A detailed procedure to determine the boundary limits of dilemma zone is explained below.

   a. **Classification of speed bands**: First the approaching speeds of vehicles or crossing speeds of pedestrians are classified into bands.

b. **Setting 10th percentile and 90th percentile boundaries classified speed bands**: Cumulative 10th percentile minimum accepted gaps and 90th percentile maximum rejected gaps are determined for various speed bands. After that, mark the midpoint of each speed band corresponding to the 10th percentile minimum accepted gaps and 90th percentile maximum rejected gaps.
c. **Plotting profile**: 10th and 90th percentile profiles for the midpoints are fitted and then lower and upper boundary limits of dilemma zone are taken on x-axis by projecting the average speed over the profiles.

8.4 PEDESTRIAN SURROGATE MEASURES: ARE THEY DIFFERENT

Most of the researchers used either Time-to-Collision (TTC) or Post-Encroachment Time (PET) as surrogate safety measure (SSM) to analyze and model the P-V interactions using video data. The previous studies (Amado et al., 2020) on P-V interactions were mainly concentrated on the discussion of various factors influencing their interaction severity levels. The use of PET or TTC alone is insufficient to measure the injury severity of a potential collision and the inclusion of either pedestrian or vehicle parameters will better predict the potential possible collisions (Gyimah et al., 2016).

8.4.1 Need for Pedestrian-Vehicle Interaction-Level Analysis

Intersections are one of the major causes of road accidents and around 25–30% of total road accidents in India are occurring at intersections only (MoRTH, 2023). According to the MoRTH road accident statistics 2023 report, approximately 70% of total accidents at intersections in India are occurring at uncontrolled intersections followed by signalized (10%), police controlled (8%), stop signs (6%), and flashing signal/blinker (6%) intersections. This high proportion of accidents at uncontrolled intersections in developing countries like India is mainly due to the heterogeneous traffic conditions and the difficulty in correct judgment while crossing the road due to the large variations in vehicular speeds. Three-legged and four-legged uncontrolled intersections are the major locations of road accidents and they accounted for more than 50% of total accidents at uncontrolled intersections.

Despite some fluctuations, pedestrian deaths are increasing every year (MoRTH, 2023). Intersections are one of the major pedestrian road crossing locations and they accounted for around 24% of total pedestrian deaths at intersections in the United States [National Highway Traffic Safety Administration (NHTSA), 2020 report]. This proportion may be higher in India where the traffic is heterogeneous and non-lane-based behavior is followed by the drivers. The high proportions of pedestrian deaths at intersections are because of the misunderstanding between vehicles and pedestrians due to dilemma behavior while crossing. A better understanding of P-V interactions is required to provide better pedestrian facilities at intersections to reduce safety-related issues.

In the past, researchers identified various factors affecting the severity levels of P-V interactions using accident data and SSMs. They also proposed the threshold limits for various severity levels of P-V interactions using either pedestrian or vehicle characteristics. But the severity depends on both pedestrian and vehicle characteristics. Determining the threshold limits for various P-V interaction severity levels using both pedestrian and vehicle characteristics will better correlate the actual severity in P-V interaction analysis and modeling.

8.5 MEASURING PEDESTRIAN DILEMMA ZONE ESTIMATION

The dilemma behavior of pedestrians while crossing is also one of the major reasons for P-V accidents. Misjudgment due to the dilemma behavior of pedestrians when the vehicle lies within a certain zone will cause P-V accidents. Providing the lower limit (unsafe gap) and upper limit (safe gap) for this dilemma behavior of pedestrians (known as pedestrian dilemma zone (PDZ)) will improve the gap acceptance behavior and safety.

P-V accidents are one of the major reasons for fatalities and the severity of these accidents varies with various parameters of pedestrian, vehicle, geometry, weather, and environment. An accident is a rare event and it is difficult to collect accident data during an accident. Many parameters are not possible to collect during an accident and their effect on severity is unknown. SSMs are indirect safety measures that measure the severity of near-miss traffic events. Analysis and modeling of P-V interactions using SSMs will help to know the various factors affecting the severity of a potential possible interaction.

The severities of P-V interactions can be modeled using various modeling techniques. BL models are used for categorical variables with two categories in the dependent variable. Ordinal logit (OL) and multinomial logit (MNL) models are used for categorical variables with more than two categories in the dependent variable. The OL model prefers over the MNL model when the order of the dependent variable is important. In OL, slopes are constant across categories but intercepts are different for each category due to the parallel lines assumption. In the present study, the ordinal logit model (OLM) is used to predict the probabilities of P-V interaction severity levels.

The probabilities of P-V interaction severity levels can be predicted in order logit model using the following equations,

$$\text{Logit}\,(P(Y \leq j)) = \beta_{0j} + \beta_1 X_1 + \beta_2 X_2 + \cdots + \beta_n X_n \tag{8.1}$$

$$P(Y = 0) = \frac{\exp(\alpha_0 + X\beta)}{1 + \exp(\alpha_0 + X\beta)} \tag{8.2}$$

$$P(Y = 1) = \frac{\exp(\alpha_1 + X\beta)}{1 + \exp(\alpha_1 + X\beta)} - \frac{\exp(\alpha_0 + X\beta)}{1 + \exp(\alpha_0 + X\beta)} \tag{8.3}$$

$$P(Y=2) = \frac{\exp(\alpha_2 + X\beta)}{1+\exp(\alpha_2 + X\beta)} - \frac{\exp(\alpha_1 + X\beta)}{1+\exp(\alpha_1 + X\beta)} \tag{8.4}$$

$$P(Y=3) = \frac{\exp(\alpha_3 + X\beta)}{1+\exp(\alpha_3 + X\beta)} - \frac{\exp(\alpha_2 + X\beta)}{1+\exp(\alpha_2 + X\beta)} \tag{8.5}$$

Y = Observed P-V interaction severity level on ordinal scale (j = 0, 1, 2, 3)

β_{0j} = Intercept corresponding to j^{th} severity level

$X_1, X_2, \ldots, X_n$ = Independent variables used in the model

$\beta_1, \beta_2, \ldots, \beta_n$ = Coefficients corresponding to each independent variable

Table 8.1 shows the description of the dependent and independent variables used in the analysis and separate code/number is assigned for each. The severity level, which is defined from the threshold limits for each P-V interaction sample, was taken as the dependent variable and pedestrian, vehicle, and geometric parameters were taken as the independent variables for the development of OLM.

TABLE 8.1
Description of the variables used in the regression analysis

S. No.	Variable	Category/Units/Code
1	Severity	Dependent variable: No interaction—0, Low severe interaction—1, Moderately severe interaction—2, Severe interaction—3
2	Pedestrian gender (PG)	Male—1, Female—2
3	Pedestrian age (PA)	Child—1, Young age—2, Middle age—3, Old age—4
4	Pedestrian speed (PS)	1—≤0.5 m/s, 2—0.5–1.5 m/s, 3—1.0–1.5 m/s, 4—>1.5 m/s
5	Luggage (Lug)	0—No, 1—Yes
6	Mobile usage (MU)	0—No, 1—Yes
7	Crossing type (CT)	1—Straight, 2—Rolling
8	Pedestrian platoon (PP)	1—One, 2—Two, 3—More than two
9	Vehicle type (VT)	Two-wheeler—1, Three-wheeler—2, Car—3, LCV—4, HCV—5
10	Vehicle speed (VS)	1—≤15 km/h, 2—15–30 km/h, 3—30–45 km/h, 4—>45 km/h
11	Vehicle direction (VD)	1—Through, 2—Right turn, 3—Left turn
12	Interaction location (IL)	1—Entry, 2—Exit
13	Lane distribution (LD)	1—Lane one, 2—Lane two, 3—Lane three, 4—Lane four

8.6 MODELING OF PEDESTRIAN DILEMMA ZONE (PDZ BOUNDARIES)

Logistic models can be used to model the probability of one event taking place by having the log-odds of one or more independent variables. Multinomial and ordered logit models can be used to model the categorical dependent variables of more than two categories. The BL model is preferred in the case of categorical dependent variables with two categories. A pedestrian perception of accepting the gap is a binary event where the probability of choosing the gap lies between 0 and 1. In the present case, the BL model is useful to capture pedestrian behavior in accepting and rejecting the gap to cross the road.

$$y_k = \begin{cases} 0, \text{ if } k^{\text{th}} \text{ pedestrian reject the gap} \\ 1, \text{ else} \end{cases}$$

The probability of accepting a given gap i for the pedestrian k is given by,

$$P_k(i) = \frac{1}{1+e^{-U_i}} \tag{8.6}$$

$$U_i = \alpha + \beta_1 X_1 + \beta_2 X_2 + \beta_3 X_3 + \cdots + \beta_n X_n \tag{8.7}$$

where U_i is the utility of gap i,

α is the intercept used in the model
$X_1, X_2, X_3, \ldots, X_n$ are the independent variables
$\beta_1, \beta_2, \beta_3, \ldots, \beta_n$ are the coefficients of independent variables.

Logistic loss (log loss) is used to measure the goodness of fit in the case of logistic regression. As the respondents are independent of one another, maximum likelihood is given by the product of the probabilities and log transformation is given in equation 8.3.

$$\ln L = \sum y_A \ln(P_A) + (1 - y_A)(1 - P_A) \tag{8.8}$$

Where, P_A is the probability of accepting the given gap.

The pedestrian has to decide whether the available gap is sufficient or not to cross the road. Sometimes this decision-making is difficult when the vehicles are within a certain region on the road. This region is known as the PDZ and it can be estimated by developing a binary logistic regression model between dependent (gap accept/reject) and independent variables (pedestrian age, pedestrian gender, pedestrian speed, location of pedestrian crossing, vehicle type, vehicle distance, and the vehicle approaching speed).

8.7 PEDESTRIAN SAFETY IMPROVEMENT PROVISIONS AND CODES OF PRACTICE

Ensuring P-V interaction safety requires a combination of engineering measures, enforcement strategies, and public awareness initiatives. Several national and international codes, standards, and methodologies guide pedestrian safety interventions. These include road design standards, traffic control measures, and best practices for improving pedestrian infrastructure and reducing P-V conflicts.

8.7.1 Indian Standards and Guidelines

In India, pedestrian safety measures are primarily governed by guidelines from the Indian Roads Congress (IRC) and regulations outlined in the Motor Vehicles (Amendment) Act, 2019.

8.7.1.1 IRC:103–2012—Guidelines for Pedestrian Facilities

This guideline provides standards for pedestrian-friendly infrastructure in urban and rural settings.

Key Provisions

- Sidewalk width: Minimum 1.8 m for low-density pedestrian areas and 2.5 m or more for high-density areas.
- Pedestrian crossings: At signalized intersections, pedestrian crossings should be at least 3 m wide and marked with zebra stripes.
- Grade-separated crossings: Foot over bridges and underpasses should be provided on high-speed corridors.

8.7.1.2 IRC:67–2012—Code of Practice for Road Signs

This standard prescribes road sign specifications for pedestrian safety.

Key Provisions

- Installation of "Pedestrian Crossing" (W-10) warning signs at zebra crossings.
- Use of reflective materials for pedestrian signs in low-visibility areas.
- Placement of "School Zone" signs near educational institutions.

8.7.1.3 IRC:86–2018—Geometric Design Standards for Urban Roads

Focuses on geometric improvements for pedestrian safety.

Key Provisions

- Curb radii at intersections should be minimized to reduce turning speeds.
- Pedestrian refuge islands should be provided on wide roads to facilitate safe mid-block crossings.
- Raised pedestrian crossings recommended for areas with high foot traffic.

8.7.1.4 IRC: SP:99–2013—Guidelines for Traffic Calming Measures in Urban and Rural Areas

Outlines traffic calming techniques to slow down vehicles in pedestrian-prone areas.

Key Provisions

- Use of speed humps, rumble strips, and raised crossings to reduce vehicle speeds near pedestrian crossings.
- Implementation of chicanes and curb extensions to narrow roads and force vehicles to slow down.

8.7.1.5 Motor Vehicles (Amendment) Act, 2019

Strengthened penalties for traffic violations related to pedestrian safety.

Key Provisions

- Increased fines for vehicles failing to yield to pedestrians at crossings.
- Introduced hit-and-run penalties for accidents involving pedestrians.
- Encouraged the use of intelligent traffic management systems to monitor pedestrian safety.

8.7.2 International Standards and Best Practices

Several international organizations, including the Federal Highway Administration (FHWA (2009), the American Association of State Highway and Transportation Officials (AASHTO), and the WHO, have developed guidelines for improving pedestrian safety.

8.7.2.1 Manual on Uniform Traffic Control Devices, USA

The Manual on Uniform Traffic Control Devices provides a standardized approach to traffic control devices, including pedestrian safety measures.

Key Provisions

- Crosswalk markings: Specifies the use of ladder-style or zebra crossings for improved visibility.
- Pedestrian signals: Calls for countdown timers and flashing beacons at crosswalks with heavy pedestrian movement.
- Pedestrian hybrid beacons (HAWK signals): Used at mid-block crossings to provide safe pedestrian crossings without disrupting traffic flow.

8.7.2.2 AASHTO Green Book (Geometric Design of Highways and Streets)

The AASHTO Green Book is widely used for road design and includes pedestrian safety provisions.

Key Provisions

- Minimum sight distance requirements at pedestrian crossings to ensure clear visibility for drivers.
- Recommended curb extensions and pedestrian refuge islands for safe crossing at intersections.
- Traffic signal timing adjustments to allow sufficient crossing time for pedestrians, including seniors and children.

8.7.2.3 WHO Global Status Report on Road Safety (2023)

The WHO's road safety guidelines emphasize a **safe system approach**, recognizing that human errors are inevitable, but road designs can minimize their consequences.

Key Recommendations

- Speed management: Speed limits should be below 30 km/h in pedestrian-heavy areas.
- Infrastructure enhancements: Wider sidewalks, designated pedestrian lanes, and traffic calming strategies.
- Education and enforcement: Road safety awareness campaigns targeting both pedestrians and drivers.

8.8 COMMONLY USED METHODS FOR PEDESTRIAN SAFETY

Several engineering and enforcement-based methods have been implemented worldwide to improve pedestrian safety.

8.8.1 Engineering Measures

These methods involve physical infrastructure changes to improve pedestrian safety and are briefly outlined as follows:

a. Traffic Calming Measures

- Speed humps and raised crossings to slow down vehicles near pedestrian crossing zones.
- Rumble strips at high-risk pedestrian crossing areas to alert drivers.
- Narrower lanes and curb extensions to reduce speeding.

b. Pedestrian Crossing Enhancements

- Zebra crossings with reflective paint for better night-time visibility.
- Pelican (Pedestrian Light-Controlled) Signals to provide dedicated pedestrian crossing phases.
- Pedestrian refuge islands in the middle of wide roads for safer staged crossings.

c. Grade-Separated Crossings

- Foot over bridges and underpasses at locations with high pedestrian traffic and high-speed vehicles.
- Elevated crosswalks in school zones to ensure pedestrian priority.

d. Enforcement Based

These methods involve monitoring and penalizing violations to improve compliance.

- Automated enforcement cameras at pedestrian crossings to catch rule violations.
- Fines for blocking pedestrian crosswalks at intersections.
- Strict penalties for jaywalking in high-risk areas.

8.8.2 Awareness and Education Programs

- School-based pedestrian safety education for children.
- Public campaigns on pedestrian right-of-way rules for both drivers and pedestrians.
- Community-led safety audits, to identify high-risk pedestrian areas and suggest improvements.

8.8.2.1 Case Study: Pedestrian Safety Enhancement in Chennai, India

Chennai is one of India's most densely populated cities, with high pedestrian footfall in commercial and residential areas. Uncontrolled pedestrian crossings on major roads led to frequent P-V conflicts, particularly in areas near metro stations, shopping zones, and educational institutions. A study conducted by IIT Madras found that pedestrian fatalities accounted for 40% of total road crash deaths in Chennai between 2015 and 2019.

8.8.2.1.1 Measures to Overcome the Problem

The Chennai Corporation and Chennai Traffic Police implemented the following pedestrian safety interventions:

1. Raised Pedestrian Crossings
   - Implemented at 40 high-risk locations, including major intersections and near schools.
   - Reduced vehicle speeds by 30–40% near pedestrian crossings.

2. Pelican Signals (Pedestrian Light-Controlled Signals)
   - Installed at busy mid-block crossings to allow pedestrians to stop traffic before crossing.
   - Pedestrians could press a button to activate a red signal for vehicles.

3. Pedestrian-Only Zones
   - Pilot project in Pondy Bazaar, converting certain streets into pedestrian-friendly zones with restricted vehicle movement.

Outcomes

- Forty-seven percent reduction in pedestrian crashes within the first year of implementation.
- Increased pedestrian compliance, as 74% of pedestrians used designated crossings post-implementation (compared to 43% before).
- Public satisfaction surveys showed that 80% of respondents felt safer while walking in areas where interventions were made.

8.8.2.2 Traffic Calming Measures for Pedestrian Safety in New Delhi, India

High vehicle speeds and mixed traffic conditions in Connaught Place and India Gate areas created unsafe conditions for pedestrians. Lack of marked crossings at key junctions led to jaywalking and mid-block crossings, increasing accident risks. A pedestrian movement study by the Delhi Traffic Police found that over 60% of pedestrians did not use designated crossings due to poor accessibility.

8.8.2.2.1 Measures to Overcome the Problem

A joint initiative by Delhi Traffic Police and the Unified Traffic and Transportation Infrastructure (Planning & Engineering) Centre introduced the following:

1. Traffic Calming Strategies
    - Speed tables and raised pedestrian crossings were installed at busy intersections, forcing vehicles to slow down.
    - Rumble strips and chicanes were added to narrow roads and control vehicle speeds.

2. Better Signage and Pavement Markings
    - High-visibility zebra crossings with reflective thermoplastic paint were introduced.
    - Signage indicating pedestrian right-of-way was placed at intersections.

3. Pedestrian Refuge Islands
    - Introduced on wide arterial roads to allow pedestrians to cross in two stages safely.

Outcomes

- Thirty-five percent reduction in P-V collisions at key intersections.
- Average vehicle speed reduced by 20–30% in areas with speed tables and traffic calming measures.
- Pedestrian compliance increased from 45% to 75%, as safer crossings encouraged proper usage.

8.8.2.3 Case Study: Vision Zero and Pedestrian Safety in Sweden

Sweden, despite having a strong road safety record, still faced pedestrian deaths, particularly in urban areas with mixed traffic. The Vision Zero initiative, launched in 1997, aimed to eliminate road fatalities, with pedestrian safety as a key focus.

8.8.2.3.1 Measures to Overcome the Problem

The Swedish government and local municipalities implemented the following under Vision Zero:

1. Safe System Approach
    - Road designs were adapted to forgive human errors, minimizing the impact of P-V crashes.

2. Speed Limit Reductions
    - Speed limits were reduced to 30 km/h in pedestrian-heavy zones, significantly reducing pedestrian fatality risks.

3. Raised Crosswalks and Advanced Pedestrian Signals
    - Raised crosswalks and signal timing changes were introduced to ensure pedestrians had enough time to cross safely.

4. Strict Enforcement and Public Awareness
    - High penalties for speeding and pedestrian right-of-way violations were enforced.

Outcomes

- More than 50% reduction in pedestrian fatalities in urban areas since Vision Zero's implementation.
- Compliance with pedestrian crossings increased by 75% due to better infrastructure and enforcement.
- The model was adopted by multiple countries, including Canada, the United Kingdom, and the United States.

The above case studies provide a better path to improve pedestrian safety using the following solutions:

- Infrastructure-based solutions like raised pedestrian crossings and speed tables are effective in reducing pedestrian crashes.
- Traffic calming measures such as speed limits, curb extensions, and pedestrian refuges significantly improve pedestrian safety.
- Innovative approaches like shared spaces can enhance safety without increasing traffic congestion.
- Strict enforcement and public awareness campaigns play a crucial role in improving pedestrian behavior and compliance.

8.8.3 Comments on Current Status and Scope for Improvement

P-V interaction safety remains a critical concern globally, particularly in developing countries like India, where pedestrian fatalities account for a significant portion of road traffic deaths. Despite the existence of codes, policies, and interventions, challenges such as poor enforcement, inadequate infrastructure, and mixed traffic conditions continue to threaten pedestrian safety. This section evaluates the current status of pedestrian safety measures and explores opportunities for further improvements in both urban and rural settings.

8.9 CURRENT STATUS OF PEDESTRIAN SAFETY MEASURES

Pedestrian safety continues to be a significant concern in urban planning and transportation. Many cities are implementing measures to improve safety.

- **Infrastructure and Road Design**
  - Many cities have implemented dedicated pedestrian crossings, foot over bridges, and pedestrian-only zones.
  - However, poor maintenance of pedestrian infrastructure (e.g., faded zebra crossings, broken sidewalks) reduces effectiveness.
  - Lack of universal accessibility features (e.g., ramps for wheelchairs, tactile paving for the visually impaired) in many pedestrian facilities.
  - Uncontrolled intersections remain a major hazard for pedestrians, with high-speed vehicles failing to yield the right-of-way.

- **Traffic Calming and Speed Management**
  - Speed management strategies such as raised pedestrian crossings, speed humps, and curb extensions have been implemented in select areas.
  - However, enforcement of speed limits in pedestrian-heavy zones is weak, leading to high-speed P-V crashes.
  - In mixed traffic conditions, especially in Indian cities, motorized and non-motorized road users interact in an unregulated manner, increasing conflict points.

- **Enforcement of Pedestrian Safety Laws**
  - Pedestrian right-of-way laws exist, but lack of enforcement leads to low driver compliance at pedestrian crossings.
  - In many Indian cities, jaywalking is common, often due to inadequate pedestrian infrastructure rather than disregard for safety.
  - Automated enforcement (e.g., pedestrian crossing cameras, speed cameras) is still in its early stages and not widely deployed.

- **Public Awareness and Education**
  - Many pedestrians lack awareness of safe crossing practices, and drivers often do not yield to pedestrians, even at marked crossings.
  - Educational campaigns are conducted occasionally but are not sustained for long-term behavioral change.

- **Policy and Government Initiatives**
  - Several government bodies, such as MoRTH and IRC, have issued pedestrian safety guidelines.

- The Motor Vehicles (Amendment) Act, 2019, has introduced stricter penalties for traffic violations, but enforcement remains inconsistent.
- Smart City projects in some Indian cities have prioritized pedestrian-friendly urban design, but implementation is limited to select areas.

8.9.1 Scope for Improvement in Pedestrian Safety

- **Infrastructure and Road Design Improvements**
  - Adoption of Complete Streets concept: Roads should be designed to accommodate all users, including pedestrians, cyclists, and vehicles, rather than prioritizing motorized traffic.
  - Wider and better-maintained sidewalks: Implement barrier-free footpaths with proper lighting and universal accessibility features.
  - Pedestrian refuge islands: Install mid-block pedestrian refuges on wide roads to enable safer two-stage crossing.
  - Improved pedestrian signals and countdown timers: Increase the allotted pedestrian crossing time at signalized intersections, especially for the elderly and disabled.

- **Traffic Calming and Speed Management**
  - Expand speed limit enforcement in pedestrian-heavy areas: Implement automated speed enforcement near pedestrian crossings and schools.
  - Increase adoption of raised pedestrian crossings and rumble strips: These force vehicles to slow down at high pedestrian activity zones.

- **Enhanced Enforcement and Regulations**
  - Use of technology for enforcement: Deploy CCTV cameras at pedestrian crossings to detect violations and fine offenders.
  - Stricter penalties for speeding in pedestrian-prone areas: Increase fines and implement a point-based penalty system for repeat offenders.

- **Public Awareness and Education Programs**
  - Pedestrian safety education in schools: Introduce road safety education as a mandatory subject in primary and secondary schools.
  - Driver training programs: Enforce pedestrian safety training in driving license tests to improve driver compliance with pedestrian priority rules.
  - Public campaigns on pedestrian safety: Launch long-term media campaigns (TV, radio, and social media) to educate both drivers and pedestrians on safe road behavior.

- **Integration of Pedestrian Safety into Urban Planning**
  - Incorporate pedestrian safety measures in Smart City initiatives: Cities should integrate pedestrian safety into urban mobility plans rather than treating it as an afterthought.

- Encourage transit-oriented development (TOD): Design mixed-use urban spaces where pedestrian-friendly infrastructure is prioritized around transit hubs.

8.9.2 Challenges in Implementation

P-V interaction safety relies on infrastructure improvements, traffic control measures, enforcement, and awareness programs. While India has several codes and policies addressing pedestrian safety, implementation gaps remain a major challenge. By adopting international best practices and integrating them with local traffic conditions, pedestrian safety in India can be significantly improved.

Implementing P-V interaction safety measures in India presents multiple challenges, ranging from infrastructure deficits and weak enforcement to behavioral issues and financial constraints. While several safety initiatives have been introduced, their effectiveness is often limited by poor execution, lack of coordination between agencies, and public non-compliance. This section explores the key challenges hindering the successful implementation of pedestrian safety measures in India.

8.9.3 Infrastructure-Related Challenges

- **Lack of Proper Pedestrian Infrastructure**
  - Narrow or non-existent footpaths: Many Indian cities lack continuous, well-maintained sidewalks, forcing pedestrians to walk on the road.
  - Encroachments on footpaths: Vendors, parked vehicles, and construction debris often reduce available walking space, increasing P-V conflicts.
  - Inaccessible pedestrian crossings: Poorly designed crossings make it difficult for elderly, disabled persons, and children to navigate safely.

- **Poor Maintenance of Existing Facilities**
  - Faded zebra crossings reduce driver compliance and pedestrian visibility.
  - Damaged sidewalks and overgrown vegetation force pedestrians onto the road.
  - Malfunctioning pedestrian signals discourage safe road crossing behavior.

- **Uncontrolled Intersections and Mixed Traffic Conditions**
  - Many uncontrolled intersections lack designated pedestrian crossings, leading to jaywalking and erratic pedestrian movement.
  - In mixed traffic environments, motorized and non-motorized users (pedestrians, bicycles, rickshaws, two-wheelers, and heavy vehicles) compete for space, increasing safety risks.

8.9.4 Enforcement and Legal Challenges

- **Weak Enforcement of Pedestrian Right-of-Way Laws**
  - Indian traffic laws mandate that vehicles yield to pedestrians at marked crossings, but compliance is poor due to weak enforcement.
  - Lack of penalties for drivers who fail to stop at pedestrian crossings discourages adherence to safety rules.
  - Jaywalking is common, but rather than penalizing pedestrians, authorities need to provide safer pedestrian infrastructure.

- **Ineffective Implementation of Traffic Calming Measures**
  - Measures such as speed humps, raised crossings, and road narrowing are implemented inconsistently and often ignored by drivers.
  - Lack of automated enforcement (e.g., speed cameras, red-light cameras) leads to low compliance with speed limits in pedestrian-heavy areas.

- **Coordination Issues between Agencies**
  - Multiple agencies (MoRTH, NHAI, local municipalities, traffic police) are responsible for road safety, leading to poor coordination in implementing pedestrian safety measures.
  - Urban development and traffic management authorities often prioritize vehicle movement over pedestrian needs, delaying pedestrian safety initiatives.

8.9.5 Behavioral and Cultural Challenges

- **Low Awareness among Pedestrians and Drivers**
  - Many drivers do not respect pedestrian right-of-way, particularly at uncontrolled intersections.
  - Pedestrians often ignore traffic signals and cross roads in risky locations, increasing accident risks.
  - Lack of road safety education contributes to unsafe behaviors among both drivers and pedestrians.

- **Public Resistance to Change**
  - Infrastructure changes such as pedestrian-only zones and road narrowing often face resistance from shop owners, vehicle users, and local businesses.
  - Encroachments on footpaths and pedestrian crossings are tolerated, making it difficult to reclaim pedestrian spaces.

8.9.6 Financial and Policy Challenges

- **Inadequate Budget Allocation for Pedestrian Safety**
  - Most transportation budgets prioritize road expansion and motor vehicle infrastructure, with minimal funding allocated for pedestrian facilities.

- Lack of dedicated pedestrian safety funds leads to delayed or incomplete projects.

- **Policy Gaps and Lack of Standardized Design Guidelines**
  - While IRC guidelines exist for pedestrian facilities, implementation is often inconsistent.
  - Many road projects lack pedestrian safety audits, leading to unsafe road designs.

8.9.7 Environmental and Urban Planning Challenges

- **Urbanization and High Population Density**
  - Rapid urbanization and increasing vehicle ownership are making Indian cities more car-centric, reducing pedestrian-friendly spaces.
  - High population density leads to overcrowding on sidewalks and crossings, making pedestrian movement difficult.

- **Air Pollution and Heat Exposure**
  - Pedestrians in Indian cities face extreme air pollution and heat exposure, discouraging walking as a mode of transport.
  - Lack of shaded pedestrian pathways makes walking uncomfortable and unsafe, especially in summer months.

8.9.8 Strategies to Overcome These Challenges

Improving Infrastructure Development & Maintenance

- Ensure proper pedestrian crossings at all intersections, particularly at uncontrolled locations.
- Regular maintenance and monitoring of footpaths, pedestrian signals, and crossings.
- Develop continuous, barrier-free sidewalks with proper width and lighting.

Strengthening Law Enforcement & Penalties

- Strict penalties for vehicles failing to yield at pedestrian crossings.
- Automated enforcement systems (speed cameras, pedestrian violation cameras) at high-risk locations.
- Increased deployment of traffic police at pedestrian-heavy areas to improve compliance.

Enhancing Public Awareness and Education

- Pedestrian safety education in schools and driver training programs.
- Awareness campaigns on safe pedestrian behavior (TV, social media, billboards).

Securing Higher Budget Allocations for Pedestrian Safety

- Government investment in pedestrian safety measures, including footpaths, signals, and enforcement systems.
- Integration of pedestrian safety into Smart City and urban mobility projects.

Adopting Better Urban Planning and Policy Reforms

- Complete Streets approach to prioritize pedestrian movement over vehicular speed.
- Mandatory pedestrian safety audits for all road projects.

8.9.9 Summary

P-V interaction safety remains a critical challenge in India and globally, with high pedestrian fatality rates due to inadequate infrastructure, weak enforcement, and unsafe road user behavior. Despite existing safety policies and codal provisions, their implementation remains inconsistent, leading to increased P-V conflicts, particularly at uncontrolled intersections and high-traffic urban areas.

Pedestrian safety in India faces significant challenges, primarily due to infrastructure deficits, poor enforcement, weak policies, and behavioral issues. To create safe P-V interaction, a multi-sectoral approach is needed, integrating engineering, enforcement, education, and policy reforms. Best practices from global cities should be adapted to the Indian context, ensuring that pedestrian safety is prioritized in future urban development plans.

Ensuring safe P-V interactions requires a multi-disciplinary approach, combining engineering, enforcement, education, and policy reforms. Countries that have successfully reduced pedestrian fatalities have done so through systematic interventions, such as traffic calming, automated enforcement, pedestrian-friendly urban design, and strong legislative backing.

For India, immediate improvements in pedestrian infrastructure, stricter enforcement of right-of-way laws, and increased public awareness can contribute to significant reductions in pedestrian fatalities and injuries. The integration of global best practices tailored to India's unique traffic environment can create safer, more walkable cities that prioritize pedestrians as equal road users.

Abbreviations Heavy Commercial Vehicle (HCV); Light Commercial Vehicle (LCV).

BIBLIOGRAPHY

Amado, H., Ferreira, S., Tavares, J. P., Ribeiro, P., & Freitas, E. (2020). Pedestrian–Vehicle Interaction at Unsignalized Crosswalks: A Systematic Review. Sustainability, 12(7), 2805.

AASHTO. (2018). *A Policy on Geometric Design of Highways and Streets*. American Association of State Highway and Transportation Officials, Washington, DC.

FHA (2009). *Manual on Uniform Traffic Control Devices*. Federal Highway Administration, Washington, DC.

Gazis, D., Herman, R., & Maradudin, A. (1960). The problem of the amber signal light in traffic flow Operations Research, 8(1), 112–132.

Indian Roads Congress. (2012). IRC:67–2012: Code of Practice for Road Signs. New Delhi, India.

Indian Roads Congress. (2012). IRC:103–2012: Guidelines for Pedestrian Facilities. New Delhi, India.

Indian Roads Congress. (2018). IRC:86–2018: Geometric Design Standards for Urban Roads. New Delhi, India.

Ministry of Road Transport and Highways (MoRTH). (2019). Motor Vehicles (Amendment) Act, 2019. New Delhi, India.

MoRTH. (2023). Road Accidents in India. *Ministry of Road Transport and Highways (MoRTH)*, Transport Research Wing, New Delhi.

Pawar, D. S., Kumar, V., Singh, N., Patil, G. R. (2016). Analysis of dilemma zone for pedestrians at high-speed uncontrolled midblock crossing. Transportation Research Part C: Emerging Technologies, 70, 42–52.

Amoh-Gyimah, R., Saberi, M., & Sarvi, M. (2016). Macroscopic modeling of pedestrian and bicycle crashes: A cross-comparison of estimation methods. Accident Analysis & Prevention, 93, 147–159.

Stewart, T. (2022). Overview of motor vehicle crashes in 2020 (Report No. DOT HS 813 266). National Highway Traffic Safety Administration.

World Health Organization. (2023). Global Status Report on Road Safety. Geneva, Switzerland.

9 AI and ML Applications for Surrogate Road Safety Analysis

9.1 INTRODUCTION

The use of artificial intelligence (AI) and machine learning (ML) has emerged as one of the most promising solutions in improving road safety through real-time traffic monitoring, accident prediction, driver behavior analysis, and smart traffic management. The usage of AI and ML for smart and safe highways depends on international methods and applications and their adoption for global accessibility in road traffic accidents. AI is gaining importance worldwide with advancements in computers sophistication to act intelligently and a better decision-making in complex situations. The use of AI for road safety can be one such area where the complex decision-making in the driving task can be better handled with such kind of advanced technologies.

It is the theory and development of computer systems to perform tasks normally requiring human intelligence, such as reasoning, visual perception, speech recognition, automated learning and scheduling, decision-making. AI leverages computers to mimic the problem-solving and decision-making capabilities of the human mind. ML is a branch of AI that develops algorithms imitating human way of learning from data and gradually improving prediction accuracy based on available and predicted data.

In line with these, there is a need for collection and processing of large amounts of data from connected vehicles, sensor systems, and road infrastructure. In this direction, Big Data analytics is one of the recent developments and AI plays a major role in analyzing such kinds of large data sets. One of the greatest assets that AI illustrates, regardless of the industry, is its learning ability, its capacity to recognize data trends, and its adaptation to changes and fluctuations in those trends. Through making predictions, and identifying patterns and outliers in data, AI can adjust as necessary and provide valuable insights.

The major methods of AI applications for surrogate safety analysis are listed as shown in the Table 9.1.

Several ML and AI applications have been developed in recent years to address some of the problems related to accident prevention and severity mitigation, accident modeling, accident cause investigation and evaluation, identification of risky driving behavior, automatic incident detection and monitoring, obstacle detection, etc.

DOI: 10.1201/9781003604761-9

TABLE 9.1
AI applications for surrogate safety analysis

Predictive modeling & crash pattern analysis	Predicting high-risk zones, analyzing historical crash data	Machine learning (logistic regression, random forest, neural networks)	USDOT crash models, European Road Safety Observatory
Computer vision for traffic monitoring	Real-time traffic violation detection, congestion monitoring, driver behavior analysis	OpenCV, YOLO, TensorFlow Object Detection API	AI-based CCTV & drone surveillance in smart cities
Intelligent traffic management systems	AI-based traffic signals adapting to congestion	Reinforcement Learning, SCOOT, MOVA algorithms	London & Sydney's adaptive traffic systems
Autonomous and connected vehicle safety	AI-powered Advanced Driver-Assistance Systems (ADAS)	Deep Learning, LIDAR, Computer Vision	ISO 26262, Euro NCAP safety systems
AI-driven law enforcement	Automated violation detection, license plate recognition	Deep Learning, OCR, Edge AI	Indian Intelligent Traffic Management System, AI-vehicle traffic surveillance in London
Data-informed road safety strategies	Data-driven policy decisions and infrastructure upgrades	AI-powered crash databases and predictive analytics	Vision Zero Data Tools (Sweden, US), WHO Road Safety Data

9.2 METHODS OF ML TECHNIQUES

AI and ML methodologies can be grouped into supervised and unsupervised learning methods. Supervised learning methods, such as classification and regression, are based on historical data and make use of labeled datasets (e.g., gender) in order to train algorithms that can accurately classify data or predict outcomes. On the other hand, unsupervised learning methods, such as clustering and dimension reduction, are not based on prior knowledge to analyze data with the scope to cluster unlabeled datasets, create association rules, and reduce data dimensionality.

Classification methods are mainly represented by methods such as Artificial Neural Networks, Support Vector Machine (SVM), logistic regression, decision trees, and k-nearest neighbors (KNN). Typical examples of transport safety problems that can be formulated as classification problems are driver maneuver classification, choice models, pedestrian detection, and traffic light detection. Regarding SVM, it is a computer algorithm that learns by example (supervised learning) to assign labels to observations. This is achieved by creating support vectors that are data points closer to the hyperplane, which are used to maximize the margin of the classifier. These points help build the SVM and their deletion would result to a change in the position of the hyperplane.

Regarding unsupervised learning methods, clustering is probably the most favorite methodological family. It refers to the organization of unlabeled data into similar groups called clusters using metrics that identify their common attributes. This ends up to the collection of data items that are "similar" between them, and "dissimilar" to data items in other clusters. To this end, some proximity measures are calculated, which are used as criterion functions to assess similarities. Examples of clustering algorithms are K-means, hierarchical, and DBSCAN, which are used in studies such as driver/user profiling, driving pattern recognition, or crash pattern detection.

AV and advanced driver-assistance system (ADAS): The scientific fields of Autonomous Vehicles (AV) and ADAS are probably those that rely the most on AI capabilities among all other road transportation fields. It is necessary for their functionality to use approaches that enable vehicles to understand the road environment and geometry, identify their surroundings, navigate to their destination, and teach themselves how to drive safely e.g. respect speed limits and highway code rules, maintain safe headways, lane discipline, and control.

Surrogate safety metrics are increasingly used to diagnose traffic safety problems so that action can be taken before a serious crash happens. They are typically based on identifying the occurrence and severity of traffic conflicts involving evasive actions such as braking or swerving, also known as near-misses or close calls. Using AI, large amounts of video footage and data points from roadside cameras and sensors can be analyzed to identify close calls in which a crash was narrowly avoided. Street imagery can be used to support assessment of roads' safety characteristics. This is already being done for attribution of star ratings in road assessment programs. The next generation of star rating programs will likely benefit from more frequent and broader image data collection as well as automated image analysis using computer vision. Drones and satellites can capture additional data and will play an increasing role in road safety. Such innovations will facilitate road safety investment planning as well as monitoring of results.

9.3 SVM APPLICATIONS

The severity levels of pedestrian-vehicle (P-V) interactions can be predicted using the threshold limits of surrogate safety measures (SSMs). The present section is intended to propose the threshold limits of SSMs [Post-Encroachment Time (PET) and risk indicator (RI)] using ML algorithms at uncontrolled intersections in mixed traffic conditions. The comparison of P-V interaction severity levels at three-legged and four-legged uncontrolled intersections is discussed in the following sections.

9.3.1 Support Vector Machines Algorithm for Classification

ML is one of the most popular words and it is used for a wide range of applications in recent decades, which predicts the outcomes of an event more accurately without being explicitly programmed to do so. ML algorithms can be used for both classification and regression purposes, and they use input data to predict a new output value. ML is often categorized into four approaches (i.e., supervised learning, unsupervised learning, semi-supervised learning, and reinforcement learning) based on how an algorithm learns to become more accurate in its predictions. The algorithm in supervised ML is trained with both labeled inputs and desired outputs and mostly they are used for classification, regression modeling, and ensembling purposes. Classification algorithms use the input training data to predict the likelihood of output data that will fall into one of the predetermined categories. Mostly logistic regression, Naive Bayes, KNN, decision tree, and SVM algorithms in ML are used for classification purposes.

Logistic regression is a very basic classification algorithm that predicts the outcome of an event based on one or more independent variables. To estimate discrete values from a set of independent variables, logistic regression is used. By adjusting the data to a logit function, it aids in predicting the likelihood of an event. The best fitting of this algorithm looks like S-shape. Even though it is an efficient and simple algorithm, it can't handle a larger number of categorical variables.

The Naive Bayes classifier assumes that a certain feature's presence in a class is unrelated to the presence of any other feature. When determining the likelihood of a specific result, a Naive Bayes classifier would consider each of these characteristics independently, even if these features are related to one another. It is an effective tool for large datasets and simple to construct.

KNN is a straightforward algorithm that classifies any new cases by getting the consent of at least k of its neighbors and then saves all the existing cases. KNN can be easily comprehended by using an analogy to everyday life. It is computationally expansive, and variables should be normalized.

The SVM algorithm is one of the supervised ML methods, mostly used for classification (binary and multiclass) and regression purposes. This algorithm is more effective in high-dimensional spaces where the number of samples are less than the number of dimensions. The probability estimates cannot be directly measured using this algorithm. In this method, the data points are plotted in n-dimensional space to perform classification by fitting the best hyperplane to separate the two classes with the largest margin. For non-linearly separable data kernelized SVM is used to transform the one-dimensional data into two dimensions. The transformed data in two dimensions will become linearly separable. The kernel function measures the similarity between the data points in the newly transformed space. Important parameters of kernelized support vector classifier are given below,

i. Kernel function: The kernel function is a technique for transforming data from its input form into the format needed for processing it. The term "Kernel" is used because SVM uses a collection of mathematical operations to provide the window through which the data can be manipulated. The various kernel functions are;
   a. Gaussian kernel: When there is no prior knowledge of the data, transformation is performed using the Gaussian kernel.
   b. Radial basic function kernel: It is same as the Gaussian kernel function but a basic method is added to this function to improve the transformation.

    c. Sigmoid kernel: The sigmoid kernel function, which serves as an activation function for artificial neurons, is equivalent to a two-layer perceptron model of the neural network.
    d. Polynomial kernel: It is used in image processing. It displays the similarity of the vectors in the training set of data in a feature space over the polynomials of the initial variables used.
    e. Linear kernel: It is used when the data is linearly separable.

ii. Gamma: The distance of a single training point is influenced by the gamma parameter. More points are together if the gamma parameter is low, and the points must be closed to each other in case of a higher gamma parameter. The gamma parameter is useful for anon-separable data.

iii. Cost (C) parameter: The C parameter incurs the penalty for any misclassified data. The penalty is low for misclassified data if the C is minimal and, due to the high-penalty, SVM aims to reduce the number of misclassified examples if the C is big.

9.3.2 Defining the Threshold Limits for Various Severity Levels of P-V Interactions

SVM classifiers offer good accuracy and perform faster prediction compared to other classification techniques. SVM works well with a clear margin of separation and with high-dimensional space. SVM draws a decision boundary which is a hyperplane between any two classes to separate them or classify them. In general, 70% data was used as training data for the algorithm and the remaining 30% as testing data for validation purposes. The threshold values of RI derived from SVM technique for various pedestrian and vehicle characteristics are determined using SVM technique and for more details on the results please refer to the reference Govinda et al. (2022).

9.4 CASE STUDIES

9.4.1 iRASTE Nagpur

Project iRASTE aims to re-imagine road safety with the predictive power of AI. For the first time, AI will act as a force multiplier and transform road safety engineering. Predictive insights generated via AI can prevent accidents even before they happen. Over time, the project expects to impact a 50% decline in road accidents in Nagpur city and black spot reduction on the city's road network.

a. **Vehicle Safety**: As a first step, a total of 150 buses of NMPL's fleet were equipped with ADAS systems, referred to as "primary fleet" hereafter. Additionally, a fleet of 50 school buses ("school fleet") was used to study the driving behavior of school bus drivers. During the intervention period, the AI-powered, ADAS-equipped NMC fleet was driven for more than 65 lakh kilometers, covering the various arterial, sub-arterial roads and state and national highways in the Nagpur metropolitan area. ADAS device provides four types of visual/audio alerts to drivers in cabin. These include Forward Collision Warning, Headway Monitoring Warning, Lane Departure Warning, and Pedestrian Collision Warning.

Outcomes

- From the NMC crash statistics of Jan '23–Aug '23 period, the 150 ADAS buses reported 41% lower accidents compared to 250 non-ADAS buses. This comparison was done by normalizing for a set of 50 buses.
- The driver risk scores have shown a consistent decline, and these are about 30% lower as compared to the values at the launch of the intervention.

b. Mobility Safety (Greyspots): Black spots are places where accidents and fatalities have happened in the past. Under this vector, the capability of the installed ADAS systems was tested to predict potential areas and stretches called "Greyspots" on Nagpur City's road network where built environment factors (road geometry) interact with the human factors (driving behavior, mix of traffic) to increase the likelihood of road crashes. The ADAS system installed on the buses generates huge volumes of alert data along the road network as a result of the warnings that the drivers accumulate as they deal with a range of traffic conditions when they traverse the network. When this dynamic data is combined with static data of the road network and geometry, spatial concentration of alerts data at specific place signals that there may be issues in the road environment that need immediate attention. Any corrective actions or interventions taken at Greyspots are proactive (not reactive) by nature, and hence can possibly help save lives due to accidents in future. Greyspots will serve as a leading indicator of road risk and can enable city authorities to plan or implement more safety interventions for better on-ground impact. This novel technique of Greyspot prediction and modeling is the first instance of the application on urban roads in India. As buses fitted with ADAS devices ply on the road network regularly, our analysis can calculate road risk data periodically and hence can be made up-to-date and accurate. Our real-time data-based approach is more accurate than traditional one-time surveys or other manual data collection methods and is suitable for a large-scalable deployment at multiple levels—be it for a city or state, across the country.

Outcomes

- The Greyspots are calculated dynamically based on the data generated by ADAS devices. To support NMC in making decisions for the purpose of road maintenance, a web-based dashboard was created to visualize the changing road environment conditions for a range of time windows—hours, days, and months.

c. 5th E in road safety—Encouragement: With Traffic Rewards organization, our project partner in Nagpur, a new Radio Frequency Identification (RFID)-based technology that nudges the NMC bus drivers to follow traffic rules based on the principle of "positive reinforcement." RFID scanners, installed at ten traffic signals across Nagpur, identify the buses with the Traffic Rewards custom tags once they halt at the traffic intersections following the signals. Each time the driver adheres to the traffic rules, he will receive points, which get added to his account in real time. These reward points can be redeemed and utilized to avail benefits from empaneled merchant establishments. As of January 2024, 438 NMC buses have been tagged. The data from the sensors across all the vehicles in Nagpur show that there has been a 24% improvement in signal adherence.

9.4.2 AI-Driven Road Safety Initiative—Kerala's AI Camera Project

The Kerala's AI Camera Project, launched under the Safe Kerala initiative, is a groundbreaking effort to leverage AI for improving road safety and reducing traffic-related fatalities. With an investment of Rs 232 crore, the project aims to curb reckless driving, enhance law enforcement, and create a safer road environment.

Implementation in the Kerala's AI Camera Project.

a. **Computational Photography**: AI-Powered Image and Video Analysis: A system that analyzes images and video using computational photography and AI algorithms to detect traffic violations such as helmet-less riding, triple riding on two-wheelers, seatbelt violations, mobile phone usage while driving, jumping red signals, over-speeding, dangerous driving, and illegal parking.
b. **Automated Violation Detection and Penalty Imposition**: AI cameras capture the violations in real time, so the motor vehicles department can automatically issue penalties without much human intervention and maintain uniformity in enforcing the law.
c. **Predictive Analytics for Accident Prevention**: AI models analyze historical accident data and real-time traffic patterns to identify high-risk zones and allow authorities to implement preventive measures such as better signage and increased patrolling.
d. **Scalable AI Network**: It has also installed a total of 726 AI-powered cameras and stationed them along prime highways and urban localities to raise the level of surveillance and enforcement.

Impact and Outcomes

- Reduction in Traffic Fatalities: The daily average of traffic-related deaths went from 12 to 7 deaths per day within the first four days of launching the project, for an overall reduction of accident-related deaths by 41.6%.
- Decrease in Traffic Violations: The number of recorded violations on the day of activation came down from 4.5 lakh to 2.8 lakh and then to 1.93 lakh cases on June 5.

- Improved Road User Compliance: The AI-based enforcement system has improved compliance among road users about helmet and seatbelt usage.
- Potential Long-Term Impact: If the current trend continues, the system has the potential to significantly contribute to the UN's goal of halving road accident deaths by 2030.
- The Kerala's AI Camera Project is a perfect example of how AI can revolutionize road safety by making it possible to monitor roads in real time, predict risks in advance, and enforce traffic regulations automatically. Such a significant decline in traffic fatalities and violations testifies to the effectiveness of AI-driven interventions. However, long-term success requires supplementing this initiative with sound policy decisions, public awareness efforts, and sustained investment in road safety infrastructure. With proper execution and adaptation, the project serves as a model for other states in India aiming to leverage AI for safer roads.

9.4.3 AI-Driven Traffic Accident Detection and Prediction Using Deep Learning

Road accidents remain a global challenge, causing significant human and economic losses annually. With the rise of social media, real-time traffic data collection has become more feasible. The study "Detection and Prediction of Traffic Accidents Using Deep Learning Techniques" proposes an AI-based system that leverages Twitter data, deep learning algorithms, and sentiment analysis to improve accident detection and prediction accuracy.

a. **Methodology:** The study employs a multi-step AI-based methodology to detect and predict traffic accidents:
   1. Data Collection and Preprocessing:
      - Data is gathered from Twitter using Application Programming Interfaces APIs, focusing on accident-related keywords.
      - Preprocessing steps include tokenization, stopword removal, and sentiment analysis.
   2. Feature Extraction:
      - Key features such as time, location, sentiment, emojis, and weather conditions are extracted.
      - Historical accident data is incorporated for predictive modeling.
   3. Deep Learning-Based Classification:
      - The system employs Long Short-Term Memory (LSTM) networks and Convolutional Neural Networks (CNN) to classify accident-related tweets.
      - A hybrid approach combining natural language processing (NLP) and deep learning improves classification accuracy.

4. Accident Prediction Model:
    - A deep learning model trained on labeled accident data predicts accident-prone locations and times.
    - ML techniques such as Random Forest and Gradient Boosting are also tested for performance comparison.

b. **AI Implementation in Traffic Accident Analysis:** The proposed system integrates various AI-driven methodologies to enhance accident detection.

1. Twitter-Based Data Collection:
    - Social media platforms, particularly Twitter, serve as a primary data source.
    - Accident-related tweets are mined and processed using NLP techniques.
2. Deep Learning for Classification:
    - The system applies deep learning models, including LSTM networks and CNN, to classify tweets as accident-related or non-accident-related.
    - The sentiment and emotion classifiers use textual analysis and emojis to enhance the accuracy of classification.
3. Predictive Analytics for Traffic Accidents:
    - Features such as geo-tagged locations, time of occurrence, weather conditions, and sentiment trends are extracted from tweets.
    - The deep learning model predicts accident-prone locations and times based on historical trends and real-time data.
4. Enhanced Feature Engineering:
    - The inclusion of parameters such as weather conditions (temperature, precipitation, wind speed) and accident severity improves prediction reliability.
    - A rule-matching classifier is used alongside AI models to refine the accuracy of predictions.

Impact and Outcomes

- Improved Detection Accuracy: The proposed deep learning model achieved 94% accuracy in classifying accident-related tweets, outperforming traditional methods.
- Enhanced Prediction Capabilities: The model demonstrated 8% improvement in accident prediction accuracy, reaching 97.5% accuracy compared to previous state-of-the-art models.
- Cost-Effective and Scalable Solution: Unlike traditional sensor-based accident detection systems, which are expensive to deploy and maintain, social media-based AI models offer a cost-effective and scalable approach.
- Faster Emergency Response: Real-time accident detection from tweets can help emergency services respond more quickly, reducing fatalities and injury severity.

- Comparison with Traditional Models: The proposed model outperformed existing ML techniques, achieving 97.5% accuracy compared to 95.3% in previous studies.

Challenges and Considerations

- Data Reliability: Social media data can be noisy and may contain misinformation, requiring robust filtering and validation mechanisms.
- Real-Time Processing: High-performance computing infrastructure is essential for handling large volumes of unstructured social media data.
- Ethical and Privacy Concerns: The use of public social media posts for surveillance and prediction raises concerns about user privacy and data security.

9.5 CURRENT STATUS AND SCOPE FOR IMPROVEMENT

The integration of AI and ML in road safety is steadily advancing worldwide. Various AI-powered technologies, such as ADAS, smart traffic management systems, AI-powered surveillance, and predictive analytics, are being implemented to reduce road accidents and enhance traffic safety. Governments and private sector players are investing in AI-driven solutions, as seen in initiatives like iRASTE in Nagpur and Kerala's AI Camera Project.

However, there are several areas for improvement:

1. Data Standardization: Road safety data collection needs uniformity across regions. Inconsistent data formats and incomplete datasets hinder AI models from delivering precise predictions and actionable insights.
2. Scalability of AI Solutions: While AI interventions have demonstrated positive impacts in controlled pilot projects, large-scale deployment remains a challenge due to infrastructural constraints and financial limitations.
3. Integration with Existing Systems: AI solutions should seamlessly integrate with existing traffic management systems, law enforcement databases, and emergency response networks for holistic improvements.
4. Public Awareness and Acceptance: Public acceptance of AI-based enforcement mechanisms, such as AI-driven traffic cameras and automated fines, remains a concern. Educating the public on the benefits of AI-based road safety interventions is crucial.
5. Ethical and Privacy Considerations: AI surveillance raises concerns about privacy, data protection, and ethical usage. Governments need clear regulations to ensure responsible AI deployment in road safety management.
6. Coordination between Various Departments: Effective implementation of AI-based road safety measures requires seamless coordination between traffic management authorities, law enforcement agencies, urban planning departments, and emergency response teams. A lack of coordination can lead to inefficiencies in executing AI-driven policies and response systems.

7. Inter-Departmental Data Sharing: Real-time data exchange between different departments, such as municipal corporations, police, hospitals, and transport authorities, is essential for AI models to function effectively. A centralized data-sharing framework can enhance predictive analytics and response mechanisms.

9.6 CHALLENGES IN IMPLEMENTATION, PARTICULARLY IN INDIA

Despite promising AI and ML applications in road safety, India faces unique challenges in implementing these technologies effectively.

1. Infrastructure Deficiencies: India's road infrastructure is often inadequate for AI-driven traffic solutions, with issues such as poor road conditions, lack of proper lane markings, and inadequate signage.
2. Cost of Implementation: Deploying AI-based road safety systems, including AI cameras and ADAS, requires significant investment, which can be a barrier for budget-constrained municipalities and government bodies.
3. Data Collection and Quality Issues: Reliable accident and traffic violation data are essential for AI models. However, data collection in India is often manual, inconsistent, and underreported, affecting the accuracy of AI-driven insights.
4. Legal and Regulatory Hurdles: India's regulatory landscape for AI implementation in traffic enforcement is still evolving. Concerns about AI-based fines, surveillance, and privacy require clear legislative frameworks.
5. High Traffic Density and Diverse Road Users: Indian roads accommodate a mix of pedestrians, cyclists, two-wheelers, auto-rickshaws, cars, and heavy vehicles, making AI-driven traffic management more complex compared to structured road systems in developed countries.
6. Resistance to Technology Adoption: Drivers, law enforcement personnel, and the general public may be resistant to AI-driven interventions due to a lack of awareness, skepticism, or fear of job losses in traditional enforcement roles.
7. Lack of Coordination between Agencies: Different government bodies—such as traffic police, municipal corporations, road transport authorities, and urban development agencies—often work in silos, leading to fragmented implementation of AI-driven road safety measures. A unified national framework is required to streamline operations and maximize efficiency.

9.6.1 Summary

AI and ML are revolutionizing road safety by enhancing real-time traffic monitoring, accident prediction, and driver behavior analysis. Case studies such as iRASTE in Nagpur and Kerala's AI Camera Project demonstrate the potential of AI to reduce traffic fatalities and improve compliance with road safety regulations. Furthermore, AI-driven accident detection using deep learning has shown promising results in leveraging social media data for predictive analytics.

Despite these advancements, challenges such as infrastructural deficiencies, high implementation costs, data collection limitations, regulatory concerns, and lack of inter-departmental coordination hinder widespread AI adoption in India. To maximize AI's impact on road safety, stakeholders must focus on improving data quality, ensuring regulatory clarity, investing in infrastructure upgrades, and fostering public awareness.

With sustained investments, policy support, and technological innovation, AI and ML can play a pivotal role in reducing road accidents and fatalities, ultimately making roads safer for all users.

BIBLIOGRAPHY

"Detection and Prediction of Traffic Accidents Using Deep Learning Techniques" - Research Paper on AI-based Traffic Safety.

"Recent Advances in Traffic Accident Analysis and Prediction: A Comprehensive Review of Machine Learning Techniques" by Noushin Behboudi, Sobhan Moosavi, and Rajiv Ramnath.

Govinda, L., Raju, M. S. K. and Shankar, K. R. (2022). "Pedestrian-vehicle interaction severity level assessment at uncontrolled intersections using machine learning algorithms." *Safety Science*, 153, 105806, ISSN 0925-7535, https://doi.org/10.1016/j.ssci.2022.105806. iRASTE Project Nagpur - AI for Road Safety, NITI Aayog.

Kerala Safe AI Camera Project, Government of Kerala.

Ministry of Road Transport and Highways, India - Road Safety Data Reports.

Nascimento, A. M., Vismari, L. F., Molina, C. B. S. T., Cugnasca, P. S., Camargo, J. B., Almeida, J. R. d., Inam, R., Fersman, E., Marquezini, M. V., and Hata, A. Y., (2020). "A systematic literature review about the impact of artificial intelligence on autonomous vehicle safety." *IEEE Transactions on. Intelligence. Transportation. Systems*, 21 (12), 4928–4946.

Policy Circle, (2024, January 30), "'How India can improve road safety with AI and data analytics.'", *Policy Circle*, 30 January. Available at: https://www.policycircle.org/society/india-road-safety-with-ai/

Road Safety in India: Status Report 2023, IIT Delhi's Transportation Research and Injury Prevention Centre (TRIP Centre).

World Health Organization (WHO) Global Status Report on Road Safety 2023.

10 Augmented and Virtual Reality in Road Safety

10.1 INTRODUCTION TO AUGMENTED REALITY AND VIRTUAL REALITY

10.1.1 What Is Augmented Reality?

Augmented reality (AR) indicates the technology that superimposes digital information (pictures, text, audio, or 3D objects) onto the real-world environment in real time. In contrast to virtual reality (VR), AR does not supplant the user's environment but augments it by projecting pertinent information. In transportation safety, AR is frequently used into:

- **Head-Up Displays (HUDs)**: Project navigation arrows, hazard warnings, and speed limits onto the vehicle windshield.
- **Smartphone Applications**: Overlay traffic alerts or construction warnings on live camera views.
- **Smart Glasses/Helmets**: Used in industries and for motorcyclists to highlight risks and provide turn-by-turn navigation (Figure 10.1).

FIGURE 10.1 Windshield with augmented reality

Source: https://www.cnbc.com/2017/11/28/wayray-uses-augmented-reality-to-turn-driving-into-a-video-game.html

DOI: 10.1201/9781003604761-10

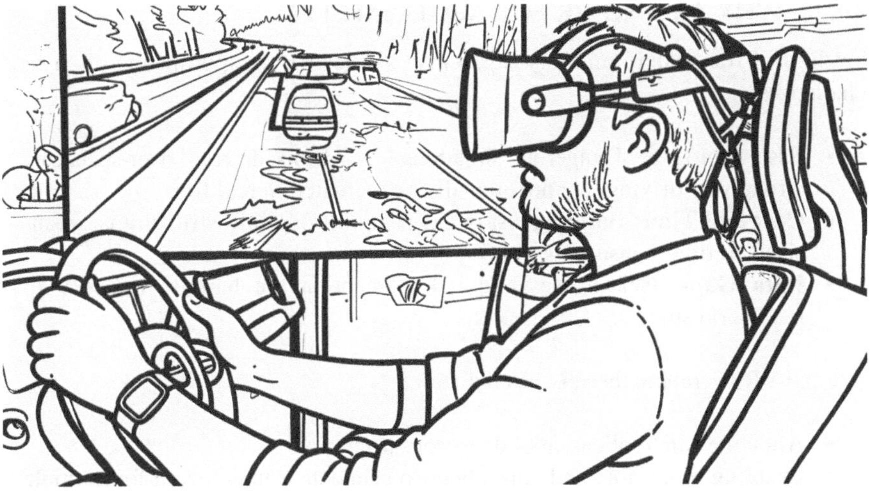

FIGURE 10.2 Virtual reality driving simulator

Source: https://cdn.prod.website-files.com/62e915854af443e54182fbc3/631d08cb84df0b1ca837a6e7_How-to-build-a-VR-Driving-Simulator-1024x536.webp

Real-World Example: Volvo introduced AR-based HUDs that project collision warnings onto the driver's field of view, such as highlighting a pedestrian about to cross, allowing faster reaction compared to relying solely on visual detection.

10.1.2 What Is Virtual Reality?

VR is a completely immersive, computer-generated environment that substitutes the real world with a simulation. VR enables academics, drivers, or pedestrians to simulate traffic conditions without real-world exposure to danger. Participants may engage with three-dimensional environments utilizing:

- **Head-Mounted Displays (HMDs)**: Like Oculus Rift or HTC Vive.
- **Cave Automatic Virtual Environment (CAVE) Systems**: A room-sized setup projecting 360° traffic scenarios.
- **VR Driving Simulators**: Equipped with steering wheels, pedals, and motion platforms for realistic training (Figure 10.2).

Real-World Example: The US National Highway Traffic Safety Administration (NHTSA) uses VR to replicate low-visibility crash scenarios (like foggy highways) to study driver responses without putting them in danger.

10.2 WHY AR AND VR FOR ROAD SAFETY?

Traditional road safety analysis (e.g., crash studies, field experiments) often faces challenges:

- **Risk Exposure**: Dangerous situations (e.g., drunk driving, over-speeding, wrong-way driving) cannot ethically be recreated in real life.
- **Cost and Time**: Building test tracks or modifying infrastructure is expensive and time-consuming.
- **Data Gaps**: Near-misses and risk perception are hard to quantify in real-world studies.

AR and VR overcome these issues by:

- Allowing safe replication of dangerous scenarios.
- Enabling repeatable and controlled experiments (same hazard for multiple drivers).
- Providing high-resolution behavioral data (eye tracking, reaction time, vehicle control inputs).

10.2.1 Illustrative Example

Imagine a busy multilane highway with limited visibility at pedestrian crossings.

- In the real world, exposing drivers to sudden pedestrian intrusions would be unsafe.
- In VR, the same scenario can be safely replicated multiple times with different pedestrians (age, walking speed), lighting conditions (day/night), and traffic flow levels.
- This provides researchers with valuable insights into driver braking response, lane-changing patterns, and pedestrian gap acceptance behavior.

10.3 APPLICATIONS OF AR AND VR FOR ROAD SAFETY ANALYSIS

The integration of AR and VR into transportation research has revolutionized the way we understand, evaluate, and improve road safety. Unlike traditional crash-data analysis or roadside observations, these technologies provide controlled, replicable, and immersive environments where risk factors can be studied without exposing participants to real danger.

10.3.1 Traffic Conflict and Crash Scenario Simulation (VR)

10.3.1.1 Application

VR allows researchers to replicate high-risk driving conditions that cannot be ethically or practically duplicated in reality. For instance, driving under the influence, exceeding speed limits in inclement weather, or navigating motorways in the incorrect direction can be analyzed without jeopardizing participants' safety (Figure 10.3).

FIGURE 10.3 Crash scenario simulation

Source: https://images.squarespace-cdn.com/content/v1/567064170e4c110e42b13b3e/1549989814086-AYI4F1V056QZKXXFRQO2/thumbnail1.jpg?format=1000w

Real-World Example

- The University of Iowa National Advanced Driving Simulator (NADS) employs a large-scale VR simulator with a full-motion vehicle cabin to study crash-prone behaviors, such as texting while driving. The data gathered has been used to develop stricter mobile phone usage laws.

10.3.2 Pedestrian and Vulnerable Road User (VR and AR)

10.3.2.1 Application

- VR allows researchers to replicate pedestrian crossings and bicycle interactions with traffic, enabling the study of gap acceptance, waiting times, and risk perception.
- AR can help pedestrians or cyclists by overlaying real-time safety cues in the environment (e.g., a flashing overlay warning of oncoming vehicles at blind corners).

Real-World Example

- In the United Kingdom, VR pedestrian simulators have been used to train school children in safe street-crossing habits. Children practiced in a VR crosswalk environment, which significantly reduced unsafe crossings when tested in real life (Figure 10.4).

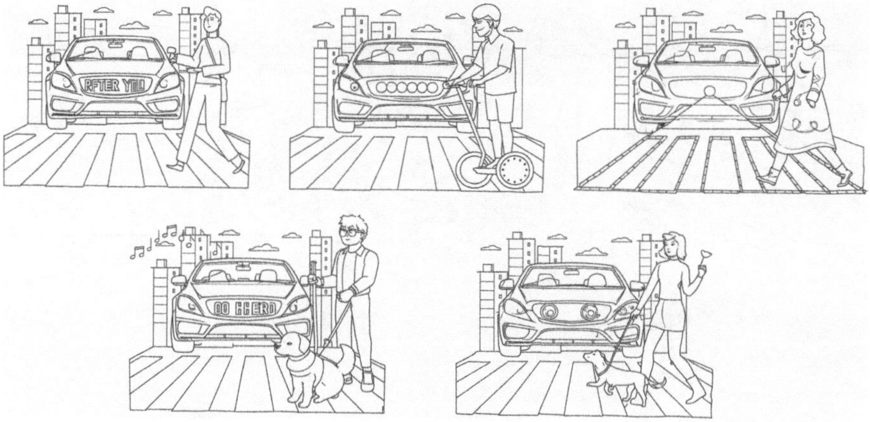

FIGURE 10.4 Vulnerable road users and connected autonomous vehicle interaction

Source: https://pub.mdpi-res.com/sensors/sensors-22-04614/article_deploy/html/images/sensors-22-04614-g008.png?1655799312

10.3.3 Road Infrastructure and Design Evaluation (VR)

10.3.3.1 Application

Prior to developing new road designs (e.g., roundabouts, pedestrian overpasses, diverging diamond interchanges), VR can evaluate driver understanding and safety results. Engineers can evaluate several designs for visibility, lane-change issues, and driver effort.

Real-World Example

- In Canada, VR was used to test driver understanding of new left-turn lane designs. Participants navigated different VR scenarios, and researchers identified confusion points before physical implementation, saving millions in redesign costs.

10.3.4 Driver Behavior Monitoring and Distraction Studies (VR and AR)

10.3.4.1 Application

- VR: Test how drivers respond to distractions (e.g., phone calls, billboard advertisements, in-vehicle infotainment).
- AR: Deliver real-time feedback to drivers about their behavior (e.g., highlighting unsafe= following distance).

Real-World Example

- Toyota developed an AR-based distraction warning system that overlays cues directly onto the windshield if a driver looks away from the road for too long.

10.3.5 Emergency and Hazardous Event Replication (VR)

10.3.5.1 Application

VR can replicate rare but high-severity events such as tire blowouts, sudden animal crossings, or brake failures. These scenarios are nearly impossible to study in field conditions but are critical for driver training and vehicle safety system evaluation.

Real-World Example

- German researchers simulated black ice conditions in VR driving simulators to analyze vehicle stability systems Anti-lock Braking System (ABS), Electronic Stability Program (ESP). The findings informed updates to driver training protocols for winter driving.

10.3.6 Connected and Autonomous Vehicles (AR and VR)

10.3.6.1 Application

AR and VR provide controlled environments to evaluate human interaction with automated driving systems. Researchers can study how drivers transition between manual and autonomous control, or how pedestrians react to autonomous vehicles at crossings.

Real-World Example

- Waymo and Uber have used VR simulations of autonomous shuttles in mixed traffic to predict pedestrian gap acceptance and driver trust levels before deploying real prototypes.

10.3.7 Illustrative Example for Road Safety Analysis

Imagine a proposed expressway bypass through a semi-urban village where mid-block pedestrian crossings are common.

- Engineers utilize VR to replicate nocturnal settings with diverse pedestrian-crossing behaviors in order to analyze driver reaction times and stopping distances.
- Utilizing AR, prospective solutions like intelligent pedestrian beacons displayed in driver HUDs can be evaluated for their efficacy in decreasing

missed detections. This combination offers diagnostic evaluation (assessing risks) and prescriptive insights (evaluating remedies).

10.4 METHODS OF AR AND VR

The effective utilization of AR and VR in road safety studies is significantly contingent upon the methodologies employed for implementation. These methods can be classified into hardware systems, software/simulation engines, data acquisition techniques, and integration strategies.

10.4.1 Hardware for AR and VR

1. **AR Hardware**
    - **HUDs**: Project data (e.g., speed, navigation, warnings) directly on the windshield without requiring the driver to look away.
    - **AR Glasses/Helmets**: Provide overlays of digital cues in real environments—used by motorcyclists and in work zones.
    - **Mobile Devices/Tablets**: Smartphones running AR apps that project road hazard warnings through the live camera view.

Example: BMW's HUD with AR overlays highlights lane departure warnings and adaptive cruise control distances on the windshield.

2. **VR Hardware**
    - **Head-Mounted Displays**: Devices like Oculus Rift, HTC Vive, or Meta Quest create immersive driving and pedestrian environments.
    - **Full-Motion Driving Simulators**: Real vehicle cabins mounted on motion platforms with 360° projection, steering wheels, pedals, and gearboxes.
    - **CAVE Systems**: Rooms with wall-to-wall projections allowing multiple participants to experience VR together.

Example: The NADS (Iowa) uses a motion platform VR setup to simulate realistic driving behavior in highway conditions.

10.4.2 Software and Simulation Engines

- **Traffic Simulation Platforms Integrated with VR/AR**
    - **VISSIM with VR plug-ins**: Simulate microscopic traffic flow and feed into VR for driver-in-the-loop experiments.
    - **Simulation of Urban Mobility (SUMO)**: Open-source platform integrated with VR for real-time traffic environment creation.

- **Game Engines for Realism**
    - Unity 3D and Unreal Engine are widely used to create realistic road networks, weather, and vehicle dynamics.
    - They allow the import of Geographic Information System data for real-city replications.

- **Safety Analysis Tools**
  - Integration with Surrogate Safety Assessment Model enables conflict analysis (Time-to-Collision, Post-Encroachment Time) in VR scenarios.

Example: Researchers in the Netherlands have developed a novel VR bicycle simulator that employs Unity 3D and VISSIM software to analyze cyclist-vehicle interactions at roundabouts successfully. This sophisticated simulator seeks to augment safety and optimize traffic flow by analyzing bicycles' navigation through intricate traffic scenarios in realistic virtual settings.

10.4.3 Human Factor Measurement Methods

To make AR/VR road safety analysis meaningful, driver/pedestrian responses must be recorded and quantified:

- **Eye Tracking**: Captures where the driver/pedestrian is looking—useful for detecting hazard perception delays.
- **Reaction Time Logging**: Measures time between stimulus (e.g., pedestrian appears) and response (brake or swerve).
- **Physiological Sensors**: Heart rate, galvanic skin response (sweating), and EEG (brain signals)—indicators of stress or workload.
- **Vehicle Dynamics Data**: Speed, acceleration, steering angle, and braking force captured directly from simulator hardware.

Example: In a VR fog-driving experiment in Japan, drivers equipped with EEG sensors showed higher brain activity in risk anticipation compared to clear-weather driving, revealing subconscious workload effects.

10.4.4 AR/VR Experimental Designs

- **Controlled Experiments**: Same hazard (e.g., sudden pedestrian entry) presented repeatedly under different conditions (day/night, rain/clear).
- **Comparative Studies**: Compare driver performance with AR assistance vs. without AR.
- **Calibration Studies**: VR outcomes compared against real-world data to ensure ecological validity.

Example: A study in Sweden compared real-world roundabout driving data with VR-simulated driving data, finding that gap acceptance times in VR closely matched field observations (±0.3 s).

10.4.5 Integration with Real-World Data

- **Digital Twin Models**: Use sensor data (traffic cameras, GPS, connected vehicle feeds) to recreate a real-time digital twin of the road in VR.
- **Mixed Reality**: Combination of AR and VR—real vehicle with VR roadway projection.

Example: Ford tested VR mixed-reality crash avoidance systems by placing real cars on a test track with virtual hazard overlays, reducing testing costs and risks.

10.4.6 Illustrative Example of Method Application

Suppose researchers want to study pedestrian gap acceptance at uncontrolled crossings on Indian highways:

1. **VR Hardware**: Participants wear a VR headset simulating a multilane rural highway with pedestrian crossings.
2. **Software**: SUMO generates vehicle traffic flow; Unity 3D creates immersive visuals.
3. **Human Factors**: Eye tracking ensures whether drivers notice pedestrians; braking data captures reaction time.
4. **Validation**: VR outcomes compared with observed field video data to check realism.

This holistic method ensures that both driver perception and pedestrian safety risks are analyzed in a safe yet realistic setting.

10.5 SAFETY TRAINING APPLICATIONS

A significant application of AR and VR in transportation engineering is in safety training. Conventional road safety training techniques—such as classroom lectures, on-road demonstrations, or videos—are frequently constrained by insufficient engagement, limited risk exposure, and the inability to replicate infrequent hazardous scenarios. AR/VR mitigates these constraints by immersing users in interactive, regulated, and replicable worlds, enabling them to practice safety skills without real-world hazards.

10.5.1 Why AR/VR for Safety Training?

- **Safe exposure to hazardous scenarios**: Participants can experience events like sudden pedestrian crossings, tire blowouts, or multi-vehicle crashes without danger.
- **Repeatability**: Same hazard can be replayed multiple times to reinforce learning.
- **Immersion and engagement**: Trainees learn better when actively participating, compared to passive observation.
- **Performance feedback**: Reaction times, errors, and decision-making can be logged for assessment.

Example: A VR truck-driving simulator allows trainees to practice brake-failure scenarios at high speeds—impossible to safely replicate in real life.

10.5.2 Applications in Road User Categories

1. **Driver Training**
    - **Novice Driver Training**: VR simulators replicate urban driving, lane changing, and merging—preparing drivers before they enter real roads.
    - **Commercial Driver Safety**: Truck and bus drivers can be trained in fatigue management, braking on gradients, and blind spot awareness.
    - **Advanced Hazard Perception Training**: Drivers encounter sudden hazards (child running across road, debris on highway) to improve anticipation.

Example: The **UK Driving Standards Agency** incorporated VR hazard perception tests into licensing, where trainees watch simulated traffic and indicate when they spot risks.

2. **Pedestrian Safety Training**
    - Children and elderly pedestrians can be trained using VR crossing environments.
    - They practice identifying safe gaps in traffic without real-world exposure.
    - AR crosswalks project digital signals (like countdown timers) into real streets to aid decision-making.

Example: The University of Alabama developed a VR street-crossing simulator for children, showing that training significantly improved gap acceptance and reduced risky crossings.

3. **Cyclist and Motorcyclist Training**
    - Cyclists can practice negotiating intersections, roundabouts, and truck blind spots in VR.
    - Motorcyclists can train for wet-surface braking, curve negotiation, and collision avoidance.

Example: The Swedish National Road and Transport Research Institute (VTI) created a VR bicycle simulator to study cyclist safety in mixed traffic, later adapted for safety training.

4. **Road Workers and Emergency Responders**
    - Work-zone workers trained in AR/VR learn to identify hazards (e.g., distracted drivers entering work zones).
    - Emergency responders practice crash rescues, hazardous material spills, and mass casualty situations in VR.

Example: In the United States, state Department of Transportation (DOTs) use VR training modules for highway maintenance crews, exposing them to near-miss work zone incidents virtually to prepare for real hazards.

Training Modules and Techniques

- **Scenario-Based Training**: Predefined dangerous events (e.g., tire blowout at 100 km/h) are introduced.
- **Gamification**: Points or rewards for safe driving decisions to improve engagement.
- **Adaptive Training**: System increases difficulty (rain, fog, nighttime) based on participant performance.
- **Collaborative VR Training**: Multiple participants (drivers, cyclists, pedestrians) interact in the same VR environment for multi-user safety training.

Evidence of Effectiveness

- Studies show that VR hazard perception training reduces crash risk among novice drivers by up to 25% in real-world driving (Agrawal et al 2018).
- Pedestrian VR training has been shown to improve safe gap acceptance behavior by 20–30% in children compared to traditional training.
- Commercial fleet operators using VR-based driver training report reduced insurance claims and lower crash frequencies.

10.5.3 Illustrative Example

Imagine training novice bus drivers in Hyderabad for pedestrian safety at rural highway villages:

1. VR simulator replicates a four-lane highway with mid-block pedestrian crossings.
2. Drivers experience villagers suddenly crossing with livestock—a frequent real-world hazard.
3. Reaction times and braking distance are recorded.
4. Performance report generated: drivers who delayed braking are flagged for additional training.

This safe yet realistic VR module allows training under culturally and context-specific conditions, something global driving simulators often miss.

10.6 DRIVER ASSESSMENT

Driver evaluation is an essential component of road safety management, as human error is constantly recognized as the predominant factor in over 90% of global road accidents (WHO, 2023). Historically, driver evaluation depended on practical driving assessments, written examinations, or monitored driving experiences. Nonetheless, these methodologies are constrained by multiple restrictions:

- Limited exposure to rare hazards (e.g., sudden tire blowouts, aggressive merging traffic).

- Variability in testing conditions (weather, traffic density, time of day).
- Subjective evaluation depending on the instructor or examiner.

AR and VR provide solutions by offering standardized, controlled, and repeatable environments to objectively measure driver behavior, decision-making, and reactions in high-risk scenarios.

Objectives of Driver Assessment with AR/VR

- Evaluate cognitive abilities: attention span, hazard perception, reaction time.
- Assess psychological and behavioral traits: aggression, compliance with rules, distracted driving tendencies.
- Monitor operational skills: steering, braking, lane discipline, and speed control.
- Identify fitness-to-drive for special groups: elderly, novice, or commercial drivers.

10.6.1 Key Parameters Assessed in VR/AR Environments

1. **Hazard Perception and Reaction Time**
   - VR scenarios simulate sudden hazards (e.g., a child running across a street).
   - The time between hazard appearance and driver response (braking, steering) is recorded.
   - This is a strong predictor of crash risk in real-world driving.

Example: A study in Australia found VR-based hazard perception scores correlate strongly with real-world crash histories in young drivers.

2. **Distraction and Attention Monitoring**
   - AR systems integrated with eye-tracking sensors monitor whether drivers check mirrors, glance at side streets, or are distracted by mobile phones.
   - VR tests simulate distractions (phone rings, billboard advertisements) to measure safety impacts.

Example: A US fleet management firm uses VR driver assessments to evaluate how delivery truck drivers handle in-vehicle distractions like incoming calls.

3. **Compliance with Traffic Rules**
   - VR-based driving tests track whether drivers obey speed limits, traffic signals, and pedestrian right-of-way.
   - Unlike real-world tests, violations are automatically logged with no examiner bias.

Example: In Singapore, VR assessment modules are included in training schools to measure how learners adapt to rules in high-density urban traffic.

4. **Risk-Taking Behavior**
   - Simulated overtaking maneuvers, aggressive lane changes, or yellow-light running are introduced.
   - System tracks if drivers choose safe vs. risky decisions.
   - Important for licensing commercial or high-capacity drivers (e.g., bus operators).

Example: In Sweden, a VR driving assessment for bus drivers tested overtaking on rural roads—identifying those prone to unsafe maneuvers under time pressure.

5. **Physical and Cognitive Fitness-to-Drive**
   - Elderly or medically at-risk drivers can be evaluated using VR environments to test reflexes, depth perception, and memory recall.
   - AR-based cognitive assessments show potential for early detection of dementia-related impairments affecting safe driving.

Example: In Japan, VR driver fitness assessments are increasingly used for elderly license renewal programs.

10.6.2 Applications of AR in Driver Assessment

- **AR Dashboards and HUDs**: Overlay performance feedback in real-time (e.g., "maintain lane discipline" warning).
- **AR Training Glasses**: Used during real driving to project visual cues and monitor gaze behavior.
- **Scenario Overlays**: Risky situations are digitally inserted into real traffic during supervised tests.

Example: BMW's AR windshields provide lane departure warnings during evaluation drives, allowing assessors to analyze driver attentiveness.

Benefits of AR/VR in Driver Assessment

- **Standardization**: Every driver faces identical test scenarios.
- **Safety**: Dangerous conditions tested virtually, not on real roads.
- **Cost Savings**: Reduces need for long-duration supervised road tests.
- **Rich Data Collection**: Eye-tracking, steering micro-movements, brake-force application.
- **Fairness**: Automated scoring reduces examiner subjectivity.

Challenges

- **Simulator Adaptation Effects**: Some drivers initially struggle with VR controls, which may bias results.
- **Motion Sickness**: VR can cause discomfort in sensitive individuals.
- **Transferability**: Need for stronger validation of how VR performance predicts real-world safety outcomes.

10.6.3 Illustrative Example

Imagine conducting driver assessment for heavy vehicle operators on India's highways:

- Drivers enter a VR simulator replicating the Delhi–Jaipur Expressway.
- Scenarios include:
  - Foggy conditions with poor visibility.
  - Sudden pedestrian intrusion at mid-block in rural village areas.
  - Work zone lane closures requiring safe merging.
- System logs reaction time, braking behavior, and lane discipline.
- A performance report categorizes drivers into low-risk, moderate-risk, and high-risk profiles.

This approach enables transport companies and regulators to effectively screen drivers before granting them permission to operate high-risk vehicles, such as trucks or buses. By implementing thorough evaluations, it can potentially lead to a significant reduction in crash rates and enhance road safety.

10.7 LIMITATIONS AND FUTURE SCOPE

While AR and VR show immense potential in enhancing road safety analysis, training, and driver assessment, these technologies are not without challenges. Understanding the limitations and identifying the future scope is essential for researchers, policymakers, and practitioners to make informed adoption decisions.

10.7.1 Limitations

1. **Cost and Accessibility**
   - High investment costs for VR simulators, motion platforms, and AR-enabled devices limit adoption, especially in low- and middle-income countries.
   - Small training institutes and government road safety agencies may find large-scale implementation economically infeasible.

   Example: A full-scale VR driving simulator with motion feedback can cost upwards of USD 100,000, which is beyond the budget of many driver schools in India and Africa.

2. **Simulator Adaptation and Learning Curve**
   - Some participants, especially elderly drivers, experience difficulty adapting to VR interfaces.
   - Training outcomes may be biased if performance reflects unfamiliarity with simulation rather than actual driving skill.

   Example: Elderly drivers in Japan often needed extra familiarization sessions before meaningful assessment could be conducted in VR tests.

3. **Motion Sickness and Health Concerns**
   - A fraction of users experience simulator sickness, including nausea, dizziness, and disorientation.
   - Prolonged use of VR headsets may cause eye strain and fatigue.
   - This limits the duration and intensity of training sessions.

4. **Validation and Transferability**
   - Although VR/AR can simulate real-world scenarios, validating transferability of performance from simulation to on-road driving remains a major challenge.
   - Some studies show strong correlations, while others reveal gaps, especially in extreme conditions (e.g., aggressive overtaking).

Example: In the United States, some VR-based driver licensing pilots showed inconsistent predictive ability for real-world crash risk.

5. **Technological Limitations**
   - VR scenarios often cannot fully replicate complex traffic dynamics, such as multi-modal interactions in Indian cities (mix of pedestrians, two-wheelers, buses, animals).
   - AR systems require precise calibration and sensor accuracy to overlay information seamlessly; errors may distract rather than assist drivers.

6. **Ethical and Privacy Concerns**
   - Continuous tracking of eye movements, reaction times, and driving behavior raises concerns about how this data is stored and used.
   - Employers or licensing authorities may misuse assessment data to unfairly classify or penalize drivers.

10.7.2 Future Scope

Despite limitations, AR and VR are expected to play an increasingly critical role in transportation safety in the coming decades.

1. **Integration with Artificial Intelligence**
   - Artificial intelligence (AI)-driven adaptive VR training will customize difficulty levels based on individual driver performance.
   - Predictive analytics can forecast which drivers are at high crash risk.

Example: A logistics company could use AI-enhanced VR training to predict which of its truck drivers are prone to fatigue-related errors.

2. **Immersive Multi-Modal Simulation**
   - Future VR platforms will integrate pedestrians, cyclists, autonomous vehicles, and shared mobility systems for holistic road safety analysis.
   - Useful for designing safer urban corridors and smart intersections.

3. **AR for Real-Time Driving Assistance**
    - AR-powered windshields and HUDs will overlay safety-critical information (pedestrian detection, collision alerts) directly in driver's field of vision.
    - Future AR could integrate with connected vehicle (V2X) systems, warning drivers of hazards beyond line of sight (e.g., hidden pedestrian behind a bus).

4. **Large-Scale Driver Licensing and Certification**
    - VR could become part of standardized driver licensing tests, reducing examiner bias and ensuring uniformity across regions.
    - Particularly valuable for commercial drivers and hazardous goods transport operators.

Example: Countries like Singapore and South Korea are already experimenting with VR-based modules in driver licensing.

5. **Low-Cost Mobile VR and AR Solutions**
    - As hardware prices fall, smartphone-based VR (Google Cardboard, Oculus Quest) and AR via low-cost smart glasses will democratize access.
    - Future apps may allow learners to practice driving scenarios at home, before formal training.

6. **Research and Policy Support**
    - Longitudinal studies are needed to validate the link between VR/AR training outcomes and real-world crash reduction.
    - Governments could establish guidelines and standards for AR/VR use in road safety, ensuring quality and ethical implementation.

10.7.3 Illustrative Example: The Future Indian Context

Imagine an AR-VR integrated national driver training program in India:

- Learners use low-cost VR headsets at home to practice hazard perception on Indian highways.
- At the driving school, they undergo VR-based risk assessments including overtaking on rural roads and navigating busy mixed-traffic intersections.
- AR smart glasses during real driving provide lane discipline cues and highlight pedestrian crossings in villages.
- The Ministry of Road Transport and Highways integrates VR performance into driver licensing systems, ensuring consistency nationwide.
- Such a system would drastically reduce the number of untrained and high-risk drivers on Indian roads. While AR and VR are not yet perfect substitutes for real-world driving exposure, they represent a transformative

tool in the pursuit of Vision Zero (eliminating road fatalities). With advancing technology, cost reductions, and robust validation studies, AR and VR will move from being innovative add-ons to becoming central pillars of road safety education, assessment, and enforcement.

BIBLIOGRAPHY

Agrawal, R., Knodler, M., Samuel, S, (2018). "Virtual Reality Headset Training: Can It Be Used to Improve Young Drivers' Latent Hazard Anticipation and Mitigation Skills." *Transportation Research Record Journal of the Transportation Research Board*, 2672 (33), 036119811875831. https://doi.org/10.1177/0361198118758311

Al-Sharif, L., and Tan, H. Z. (2019). "Virtual reality driving simulators for driver training and evaluation: A review." *Transportation Research Part F: Traffic Psychology and Behaviour*, 65, 411–425.

Allen, H. A., and Lunenfeld, H. (2018). "Driver performance assessment using virtual reality simulation environments." *Journal of Safety Research*, 67, 177–186.

Caird, J. K., Chisholm, S. L., Edwards, C. J., and Creaser, J. I. (2007). "The effect of augmented reality head-up displays on driver performance and distraction: A meta-analysis." *Transportation Research Record: Journal of the Transportation Research Board*, 2018(1), 119–128.

Chen, F., Chen, S., and Ma, X. (2018). "Crash risk evaluation of commercial drivers using virtual driving simulator data." *Accident Analysis & Prevention*, 120, 50–60.

de Winter, J. C. F., and Dodou, D. (2014). "The driver behavior questionnaire as a predictor of accidents: A meta-analysis." *Journal of Safety Research*, 49, 77–83.

Engström, J., Markkula, G., and Victor, T. (2017). "Effectiveness of driver training and assessment programs using driving simulators." *Human Factors*, 59(8), 1203–1221.

Fitch, G. M., Soccolich, S. A., Guo, F., McClafferty, J., Fang, Y., Olson, R. L., Perez, M. A., Hanowski, R. J., Hankey, J. M., and Dingus, T. A. (2013). "The impact of hand-held and hands-free cell phone use on driving performance and safety-critical event risk." *Transportation Research Record*, 2321(1), 1–13.

Hou, Y., Tang, T., and Wang, Y. (2021). "Use of immersive virtual reality in road safety education for children and young drivers: A review and future perspectives." *Transportation Research Part F: Traffic Psychology and Behaviour*, 80, 18–35.

Jamson, A. H., and Merat, N. (2005). "Surrogate safety measures for assessing the effectiveness of virtual driving training systems." *Accident Analysis & Prevention*, 37(6), 1035–1043.

Kaplan, S., Guvensan, M. A., Yavuz, A. G., and Karalurt, Y. (2017). "Driver behavior analysis for safe driving: A survey." *IEEE Transactions on Intelligent Transportation Systems*, 18(8), 2072–2084.

Kemeny, A., and Panerai, F. (2003). "Evaluating perception in driving simulation experiments." *Trends in Cognitive Sciences*, 7(1), 31–37.

Kim, H., Lee, Y., and Park, S. (2020). "Augmented reality head-up displays for advanced driver assistance systems." *IEEE Transactions on Intelligent Vehicles*, 5(1), 28–38.

Lee, J. D. (2014). "Fifty years of driving safety research." *Human Factors*, 56(6), 1036–1049.

Liang, Y., and Horrey, W. J. (2012). "Virtual reality training for hazard perception among young drivers: Transferability and limitations." *Transportation Research Part F: Traffic Psychology and Behaviour*, 15(3), 188–197.

McGill, M., Boland, D., Murray-Smith, R., and Brewster, S. (2015). "A review of virtual reality and augmented reality in road transport training and simulation." *International Journal of Human-Computer Studies*, 82, 1–13.

Regan, M. A., and Mitsopoulos-Rubens, E. (2016). "Driver distraction and inattention in the context of augmented reality." *Cognition, Technology & Work*, 18(2), 291–305.

Stanton, N. A., Salmon, P. M., Walker, G. H., and Jenkins, D. P. (2016). *Human Factors in Road and Rail Transport: Simulators and Simulation for Training, Assessment and Research.* CRC Press, Boca Raton, FL.

WHO, (2023). *Global status report on road safety 2023: Summary.* Geneva: World Health Organization; 2023. Licence: CC BY-NC-SA 3.0 IGO

Wickens, C. D., and Hollands, J. G. (2015). *Engineering Psychology and Human Performance.* Routledge, New York.

11 Automated Vehicle Safety and Future Scope

11.1 INTRODUCTION

Across the globe, the transport system is experiencing a technological transformation. For more than a century, human operators have dominated transportation; nevertheless, algorithms, sensors, and artificial intelligence (AI) are progressively assuming control. Automated vehicles (AVs) are not merely futuristic concepts; they are currently undergoing testing in urban areas, highways, and industrial zones. The impetus for autonomous vehicles is grounded in safety. Road accidents result in over one million fatalities globally each year, predominantly due to human error. By diminishing or eliminating human participation in driving activities, autonomous vehicles possess the capacity to significantly decrease fatalities and injuries. Simultaneously, automation offers enhanced traffic flow, improved energy efficiency, and increased mobility for individuals with disabilities or restricted driving capabilities.

Nonetheless, the path to the secure and dependable deployment of autonomous vehicles is not uncomplicated. In contrast to conventional vehicles, autonomous vehicles must demonstrate their capability to manage both common and infrequent traffic situations with minimum danger. Safety validation is complicated due to the infrequency of real-world crashes, rendering it unfeasible to depend exclusively on crash statistics. Surrogate safety analysis provides quantifiable, real-time metrics that indicate safety performance without the necessity of waiting for events to transpire. This chapter commences with the implementation of surrogate safety in autonomous vehicles, then addressing the existing worldwide landscape, the issues encountered, potential solutions, and future trajectories.

11.2 APPLICATION OF SURROGATE SAFETY FOR CONNECTED AND AUTOMATED VEHICLES

For decades, road safety has primarily been examined through collision reports and accident databases. This strategy is effective for conventional automobiles but inadequate for AVs. The rationale is straightforward: crashes are infrequent occurrences, and accumulating statistically relevant data may require years, if not decades. Prolonged waiting would hinder advancements in safety validation and impede the development of autonomous vehicle technologies. Researchers employ Surrogate safety measures (SSMs) to address this restriction. These indicators are obtained from near-miss incidents, vehicle trajectories, and driver interactions, offering early warning signs of probable collisions.

DOI: 10.1201/9781003604761-11

Common Surrogate Safety Measures

- **Time-to-Collision (TTC)**: The time remaining before two vehicles would collide if they continue on their current paths and speeds. A small TTC indicates high risk.
- **Post-Encroachment Time (PET)**: The time gap between one vehicle leaving a conflict point (such as an intersection) and another vehicle entering it. The shorter the gap, the higher the risk.
- **Deceleration Rate to Avoid Collision**: The level of braking effort required to avoid a crash. Higher deceleration values point to unsafe interactions.

11.2.1 Application in AVs

Connected and automated vehicles (CAVs) produce substantial data from onboard sensors, including Light Detection and Ranging (LiDAR), radar, and cameras, in addition to vehicle-to-everything (V2X) connectivity. These continuous data streams allow engineers to compute SSMs instantaneously.

- In simulations, AVs can be placed in thousands of virtual traffic scenarios, and their responses can be evaluated using SSMs.
- In controlled field trials, researchers monitor near-miss events and use SSM thresholds to assess how safely AVs navigate interactions with other vehicles, cyclists, and pedestrians.

This approach shifts the focus from waiting for accidents to actively learning from near-misses and conflict situations.

11.2.2 Illustrative Scenario

In initial testing of automated shuttles in European cities, researchers employed TTC values to analyze interactions between shuttles and people at crossings. When TTC numbers consistently dropped below safe standards, engineers enhanced the shuttle's decision-making algorithms, augmenting its capacity to halt earlier and yield safely.

11.2.3 Solution Pathway

- Define AV-specific thresholds for surrogate measures (e.g., what TTC values should trigger emergency braking in AVs).
- Apply SSMs both in virtual environments (to cover rare scenarios) and in real-world trials (to validate practicality).
- Integrate SSM results into AV control systems so that algorithms continuously improve safety performance.

By applying surrogate safety analysis, the industry can move closer to proving that AVs are not only functional but also safe enough for large-scale deployment.

11.3 CURRENT SCENARIO

The advancement of CAVs is occurring consistently, albeit at varying rates worldwide. Although complete automation is an aspirational objective, varying degrees of automation are currently seen in consumer markets and controlled experiments.

11.3.1 Technology and Deployment

Currently, numerous vehicles feature partial automation features including adaptive cruise control, lane-keeping assistance, blind spot detection, and automated parking. These features alleviate driver workload and enhance road safety by diminishing human mistake in everyday actions. In addition to consumer vehicles, autonomous taxis, shuttles, and delivery robots are undergoing trials in real-world environments. Most deployments occur in geo-fenced regions, like low-speed urban corridors or designated highway zones, where operation circumstances are foreseeable and regulated.

11.3.2 Policy and Regulation

Governments are adopting a cautious but forward-looking stance. For example:

- In Germany, drivers can legally disengage from active control under certified Level 3 systems during traffic jams on highways.
- In the United States, pilot programs for autonomous ride-hailing and delivery are authorized in states such as Arizona and California.
- In Japan, driverless delivery pods are being tested to support logistics in aging rural communities.

These examples reflect a structured progression where legal permissions expand gradually as technology proves reliable.

11.3.3 Industry Efforts

The autonomous vehicle ecosystem relies on cooperation among automotive manufacturers, software engineers, and telecommunications providers. Prominent entities are investing in LiDAR, radar, and camera-based vision technologies in conjunction with AI-driven decision-making frameworks. Telecommunication providers facilitate these systems using 5G-enabled V2X networks, enabling autonomous vehicles to interact in real time with road infrastructure, pedestrians, and other vehicles.

11.3.4 Public Perception

Despite technical advancements, trust remains a major hurdle. Surveys show that while people appreciate driver-assistance features, they are hesitant to ride in fully driverless vehicles due to safety concerns. High-profile trial incidents have amplified skepticism, underscoring the need for transparent safety validation and public engagement.

11.3.5 Illustrative Example

Germany's approval of Level 3 automation in limited scenarios demonstrates how regulators, manufacturers, and society are working in tandem. Allowing drivers to legally take their hands off the wheel in traffic jams is a symbolic but crucial step showing that automation can be responsibly introduced without overwhelming public trust.

11.4 CHALLENGES AND LIMITATIONS

While CAVs have made remarkable progress, their large-scale deployment still faces significant hurdles. These challenges extend beyond technology to include legal, ethical, social, and infrastructural concerns.

11.4.1 Technical Challenges

AVs rely on sensors, maps, and AI decision-making. However, these systems are not infallible:

- **Sensor Limitations**: Cameras and LiDARs may perform poorly in fog, heavy rain, or snow, reducing vehicle perception reliability.
- **Edge Cases**: AI struggles with unusual or rare events such as sudden pedestrian movements, erratic cyclists, or animals crossing the road.
- **Maps and Localization**: High-definition maps, essential for AV navigation, require constant updates. Even small environmental changes (e.g., roadworks, new signage) can create navigation errors.

11.4.2 Cybersecurity and Data Privacy

As CAVs become more connected, they are also vulnerable to digital threats:

- **System Attacks**: GPS spoofing, signal jamming, or hacking into vehicle controls could cause accidents or traffic chaos.
- **Data Security**: AVs collect enormous amounts of personal and travel data. Protecting this information while complying with privacy regulations remains a major concern.

11.4.3 Legal and Ethical Questions

One of the most debated aspects of AVs is responsibility and morality:

- **Liability**: If an AV is involved in a crash, determining who is at fault, the manufacturer, the software provider, or the passenger, remains unresolved.
- **Ethical Dilemmas**: In unavoidable crash situations, should an AV prioritize passenger safety or that of pedestrians? These "trolley problem"-like scenarios remain controversial and unsolved.

11.4.4 Social and Economic Impacts

Automation brings societal disruption alongside technical innovation:

- **Job Displacement**: Professional drivers in trucking, taxis, and delivery services may lose employment opportunities.
- **Public Trust**: Despite potential safety benefits, widely reported accidents involving AVs fuel public skepticism. This slows adoption even when evidence suggests improvements in overall safety.

11.4.5 Validation Gaps

The ultimate test for AVs is proving that they are consistently safer than human drivers. This remains difficult because:

- Real-world crashes are statistically rare, making direct comparisons difficult.
- SSMs offer insights but still require stronger calibration to accurately reflect real-world crash risks.

11.4.6 Illustrative Example

In 2018, a test AV in Arizona struck and killed a pedestrian during night-time trials. This incident triggered global debate over the readiness of AV systems, the adequacy of human safety drivers, and the role of regulators in protecting the public. It highlighted that even a single high-profile failure can overshadow years of progress and erode public confidence.

Despite these limitations, addressing them systematically opens pathways toward safer, more reliable automation. Each challenge carries within it an opportunity for innovation, collaboration, and new policy frameworks.

11.5 PROPOSED SOLUTIONS

The challenges identified in AV safety are complex but not insurmountable. A balanced combination of technological innovation, regulatory clarity, and public engagement can pave the way for safer adoption.

Technological Enhancements

- **Sensor Fusion**: Combining LiDAR, radar, cameras, and ultrasonic sensors minimizes the weaknesses of any single sensor. For instance, radar can perform well in fog, while cameras capture visual detail in clear conditions.
- **AI Training with Rare Events**: Using large-scale simulation platforms to expose AV algorithms to unusual and critical events (e.g., sudden pedestrian crossings, debris on the road) ensures better preparedness.

- **Real-Time SSM Integration**: Embedding SSMs directly into AV systems allows continuous monitoring of near-misses. This transforms SSMs from research tools into operational safety features.

Cybersecurity and Data Protection

- **Encryption and Authentication**: Secure communication protocols for V2X data exchange reduce hacking risks.
- **Continuous Security Audits**: Regular updates and penetration testing ensure system resilience.
- **Privacy by Design**: AV systems should anonymize personal data wherever possible to build public trust.

Regulatory and Policy Measures

- **Harmonized Standards**: International collaboration can lead to shared safety benchmarks, preventing fragmented regulations.
- **Phased Certification**: Gradual approval processes (Level 2 → Level 3 → Level 4) based on performance in surrogate safety evaluations.
- **Legal Clarity**: Clear liability frameworks ensure accountability when accidents occur.

Social and Economic Adaptation

- **Public Awareness Campaigns**: Transparency in AV performance (publishing safety statistics) can build trust.
- **Workforce Transition Programs**: Governments and companies should retrain professional drivers for roles in AV supervision, fleet maintenance, or logistics management.
- **Community-Cantered Pilots**: Involving local communities in AV trials ensures their concerns are addressed early.

11.6 FUTURE SCOPE

Looking ahead, AVs will not simply replace human drivers; they will reshape the mobility ecosystem. Their future scope can be seen across several dimensions.

Safety Evolution

- **Proactive Safety Systems**: Instead of reacting to accidents, AVs will predict and prevent them. Surrogate measures will evolve into real-time safety dashboards, constantly adjusting behavior.
- **Integration with Smart Infrastructure**: Traffic signals, pedestrian crossings, and road sensors will directly communicate with AVs, enhancing anticipation of risks.

Technological Pathways

- **AI with Explain Ability**: Future AVs will incorporate explainable AI, making their decisions transparent to regulators and the public.
- **Edge and Cloud Synergy**: Processing safety-critical data on-vehicle (edge) while using cloud platforms for learning from fleet-wide experience.
- **5G and Beyond**: Ultra-low-latency networks will allow near-instant vehicle-to-vehicle and vehicle-to-infrastructure communication.

Policy and Regulation Outlook

- **Global Safety Frameworks**: By the next decade, international treaties may standardize AV safety testing, much like aviation standards today.
- **Insurance Transformation**: Insurance models will shift from individual drivers to shared liability between manufacturers, software providers, and fleet operators.

Societal and Economic Impacts

- **Inclusive Mobility**: AVs can empower elderly individuals, people with disabilities, and rural communities with reliable transport.
- **Urban Redesign**: Cities may redesign roads, parking spaces, and traffic systems to prioritize shared AV fleets over private ownership.

11.7 CASE STUDIES IN AUTOMATED VEHICLE SAFETY

Examining real-world examples of AVs in operation helps to understand how theory translates into practice. These case studies show both the potential and the challenges that come with deploying AVs in everyday environments.

11.7.1 Case Study 1: Waymo's Driverless Taxi Service (USA)

Waymo has been running driverless taxi services in Phoenix, Arizona. Their vehicles operate without a human driver and depend on LiDAR, radar, and high-definition maps. Data suggests their crash rates are lower than those of human drivers. However, issues like hesitation at busy intersections or confusion near construction sites highlight that AVs are still adapting to unpredictable environments.

11.7.2 Case Study 2: Nuro's Autonomous Delivery Pods (Japan and USA)

Nuro's small, low-speed delivery pods transport groceries and parcels in suburban areas. Their design of compact size and capped speed reduces risks. By starting in predictable environments, they prove how phased introduction of AVs can build safety confidence before scaling to complex city traffic.

11.7.3 Case Study 3: Tesla's Autopilot and Full Self-Driving (FSD) Controversies

Tesla's semi-automated features have reduced certain crash types but have also faced criticism. Misuse by drivers and high-profile accidents have drawn global attention. This case highlights the importance of user responsibility, clear communication, and regulatory oversight in automation.

11.7.4 Comparative Global Approaches

Different regions approach AV development with their own priorities, shaped by culture, economics, and regulation.

- **United States**: Industry-led innovation dominates, supported by flexible state-level rules. While this encourages experimentation, it also creates fragmented regulations across states.
- **Europe**: Focuses on harmonized safety and consumer protection. Deployment is cautious but steady, ensuring public confidence.
- **China**: Rapidly pilots AVs in urban centers with strong government backing, aiming to lead globally through scale and smart city integration.
- **Japan**: Uses AVs to address demographic challenges, especially mobility needs of its aging population.

11.7.5 Ethical and Societal Perspectives

AVs raise ethical questions that go beyond technical reliability.

- **The Trolley Dilemma**: If a crash is unavoidable, should an AV protect its passengers or pedestrians? There is no universal answer, and public opinion varies.
- **Fairness and Access**: AVs could reduce mobility gaps if deployed as shared services. However, if they remain luxury products, they may deepen inequality.
- **Impact on Cities**: AVs may reduce private car ownership, creating space for pedestrians and green zones. On the other hand, they could encourage urban sprawl if long-distance commuting becomes cheaper.

11.7.6 Integration with Smart Infrastructure

AVs will work best when paired with intelligent transport systems.

- **Smart Traffic Lights**: Signals that communicate with AVs to reduce delays and conflicts.
- **Digital Roadside Units**: Sensors that share environmental and traffic data with vehicles.
- **Dynamic Road Pricing**: Real-time toll adjustments to control congestion.

11.7.7 Research Directions and Academic Contributions

Academic and research institutions are playing a major role in AV safety.

- **Simulation Platforms**: Universities build digital city models to run millions of virtual driving scenarios, saving cost and time.
- **Human Interaction Studies**: Research on driver–automation handover improves system design for semi-autonomous vehicles.
- **Improved SSMs**: Scholars refine metrics like TTC and PET so they better reflect real-world crash likelihoods.

11.8 FUTURE CHALLENGES BEYOND TECHNOLOGY

Even with advanced AI and sensors, AV adoption will face broader issues.

- **Energy Demand**: Electrified AV fleets will stress national power grids, demanding renewable and resilient energy solutions.
- **Data Ownership**: Questions remain on who owns and controls AV-generated data—the user, the manufacturer, or governments.
- **International Use**: Freight and cross-border operations will require legal harmonization across countries, a challenge that goes beyond technology.

11.8.1 Summary

AVs represent a paradigm shift in mobility. Their safety validation cannot depend solely on rare crash data; SSMs provide a meaningful way forward. The current global scenario shows cautious but steady progress, with governments, industries, and societies experimenting with pilot projects. Still, technical, legal, and social hurdles persist.

The solutions proposed from sensor fusion to clear regulations offer pathways to overcome these challenges. Looking toward the future, AVs hold the potential not only to improve road safety but also to transform urban design, expand mobility access, and accelerate sustainability goals. The journey will not be free of setbacks, but with thoughtful application of technology, governance, and human-centered design, AVs can achieve their promise: safer roads, smarter cities, and sustainable mobility for all.

Index

Note: **Bold** page numbers refer to tables and *italic* page numbers refer to figures.

B

C

D

T

U

V

W